Voracious Science & Vulnerable Animals

ANIMAL LIVES

A series edited by Jane C. Desmond, Barbara J. King, and Kim Marra

Voracious Science & Vulnerable Animals

A PRIMATE SCIENTIST'S ETHICAL JOURNEY

·JOHN P. GLUCK·

The University of Chicago Press   *Chicago and London*

JOHN P. GLUCK is professor emeritus of psychology and a senior advisor to the president on animal research ethics and welfare at the University of New Mexico. He is also research professor of the Kennedy Institute of Ethics at Georgetown University and the coauthor of *The Human Use of Animals.*

The University of Chicago Press, Chicago 60637
The University of Chicago Press, Ltd., London
© 2016 by The University of Chicago
All rights reserved. Published 2016.
Printed in the United States of America

25 24 23 22 21 20 19 18 17 16   1 2 3 4 5

ISBN-13: 978-0-226-37565-6 (cloth)
ISBN-13: 978-0-226-37579-3 (e-book)
DOI: 10.7208/chicago/9780226375793.001.0001

Library of Congress Cataloging-in-Publication Data
Names: Gluck, John P., 1943– author.
Title: Voracious science and vulnerable animals : a primate scientist's ethical journey / John P. Gluck.
Other titles: Animal lives (University of Chicago Press)
Description: Chicago ; London : The University of Chicago Press, 2016. | Series: Animal lives
Identifiers: LCCN 2016011996 | ISBN 9780226375656 (cloth : alk. paper) | ISBN 9780226375793 (e-book)
Subjects: LCSH: Primates as laboratory animals. | Animal experimentation—Moral and ethical aspects.
Classification: LLC SF407.P7 G58 2016 | DDC 636.98—dc23 LC
record available at http://lccn.loc.gov/2016011996

♾ This paper meets the requirements of ANSI/NISO Z39.48–1992 (Permanence of Paper).

To Charlene Dee McIver
My partner, my guide, and courageous lover of life

*

A smith who possesses tongs will not lift the glowing iron out of the coals with his own hands.

IMMANUEL KANT

In every heart there is a coward and procrastinator.

MARY OLIVER

Contents

Preface ix

Introduction *1*

1 * Erosion *13*

2 * Induction *64*

3 * Practice *111*

4 * Awareness *143*

5 * Realignment *177*

6 * Reconstruction *221*

7 * Protection *250*

8 * Reformation *282*

Epilogue *292*

Notes 297   Index 303

The photo gallery follows p. 176.

Preface

When I sat down at my desk on a Thursday in August to start this book in earnest, I was acutely aware of a faint but pungent odor coming from my clothing. It was an odor with which I had become quite familiar over the years; I consider it almost sweet, even though many people find it distasteful. The odor emanates from an oil secreted from the skin of an endangered southeast Asian monkey called the stump-tailed macaque (*Macaca arctoides*). Stump-tails are compact, stocky animals with reddish brown to dark brown hair and colorful red-and-black-speckled faces. They weigh about fifty to seventy-five pounds and have small two-inch tails that look like human thumbs. A small group of these monkeys, "retired" from research, were housed in my laboratory and were my continuing responsibility.

For the most part stump-tails are quite easygoing in their interactions with humans. I have been approached many times by both males and females who were interested in simply initiating bouts of mutual grooming. These monkeys stand in contrast to the rhesus monkey (*Macaca mulatta*), whose aggressiveness toward humans in laboratory situations is well known. For many years I had as many as eighty rhesus monkeys in my lab, and they instill fear to this day. The stump-tail odor can be transferred to one's clothing and exposed skin merely by being in close proximity to the animals for an hour or so.

On that day, the odor had been transferred to my street clothing through my lab coat as I helped to hold an old female stump-tail named Donna so that the campus veterinarian could administer a drug for the purpose of ending her life. Several days previously she had presented lethargic and relatively unresponsive with an obvious hemorrhagic (i.e., bloody) diarrhea. Subsequent tests showed that her kidneys had ceased to function effectively; she was suffering from end-stage renal disease, and there were no effective treatment options. After consulting with her caretakers and Kevin O'Hair, the attending veterinarian, I had decided to euthanize her.

I am using the words *I had decided* in the previous sentence very deliberately. Writing about such a decision earlier in my life as a researcher, I would have reflexively turned to the passive "it was decided." Such language served to pass responsibility on to trained technicians and veterinarians and dilute the ethical gravity of determining the fate of living beings. Now, however, I recognize that being an animal researcher is not just about theory construction, scientific methodology, and data collection, but about being the creator, purchaser, and terminator of lives. Lives that have, in the broadest sense, meaning for the creatures who are living them, independent of the goals that we as researchers have for them.

The decision about Donna had not been entirely straightforward. She lived in a small social group of three other animals and was strongly attached to Telemachus, the only remaining male. Donna and Telemachus, though involved with the others, clearly preferred one another's company almost exclusively. They were always in close physical contact, spending hours together each day grooming, eating, and sleeping together. We were concerned about how Tele would respond to Donna's loss. And we ourselves—the technicians and I—had known and worked with Donna her whole life, all thirty-two years of it. We were not strangers. However, we quickly acknowledged that her apparent suffering had to be stopped, and in the end the best choice was to proceed. We agreed to stay particularly alert to Tele's subsequent reaction. Although we did not say it out loud, we agreed to watch one another's responses as well.

Donna had remained still, barely resisting, when her long-term caretakers, Ector Estrada and Gilbert Borunda, approached her as she sat on a perch in the outdoor section of the enclosure. In a smooth, well-practiced move, Ector quickly controlled her upper arms with his hands, holding her with the minimum firmness required to ensure her compliance. It did not require much force, as these two individuals knew each other very well and a trust seemed to exist between them based upon hundreds of previous handling episodes. As he lifted her from the perch, she briefly held on to the wire mesh wall with her feet as if she knew what lay ahead. Ector did not pull her to release her grip; he simply stopped and waited. After a moment, her trust returned and she released her grip.

Ector then carried her the short distance to the treatment room and sat her down on the narrow stainless-steel table while maintaining his hold on her upper arms. I loosely held her legs until it was clear that she was relaxed, and then I released them. I stroked her head as the veterinarian filled a syringe with the euthanasia drug, a powerful narcotic with the hopeful name Sleep Away. He was new to our campus and had no previous experience with this group of monkeys. As he turned toward her, syringe in hand, he stopped for a moment, taking in her appearance. "She's beautiful," he said softly.

The utterance was totally spontaneous and seemed to surprise even him. He recovered his professional composure, proceeded to find a vein, and slowly pushed the liquid into her right arm.

She accepted the needle like the many she had experienced before during research and routine health checks, without a flinch. In seconds her head glided down to her left shoulder as her body fell limp. The light left her eyes, her lids slid partially closed, and there was a hint of looseness in her lower lip.

After a moment, we assured ourselves that she was dead, each in our own way: Kevin listened for breath and heart sounds with his stethoscope; Ector squeezed her thigh, probing for evidence of a femoral pulse; and I checked for an eye blink reflex. Gilbert stood silently, making sure that our checks were validly exercised. We looked at each other and nodded in agreement that Donna was dead. We placed her body into a red biohazard bag and then sealed it. The biohazard bag was necessary because years earlier Donna's blood had shown markers indicating that she had been infected with Herpes B virus at some point, and this virus can cause a fatal infection in humans.

I carried Donna's limp body down the stairs, to the parking lot, and then into the closed bed of my pickup truck. By the time I had driven the ten miles to the animal crematory and handed her over to the receiving technician, the deep warmth that I had felt when carrying her had mostly left her body. I filled out the biohazard form, giving her name and age, paid the seventy-five dollars in cash, and left.

When I returned to my office to write, the memory of Donna's life—and my appropriation of it—lingered in my mind as stubbornly as her odor on my clothing. This was entirely apt, since I had conceived of this book in part as a way of paying back a debt I felt I owed to Donna and dozens of stump-tails, rhesus monkeys, and other completely vulnerable animals. During my research career, I had caused significant harm to these animals for the sake of advancing scientific knowledge and my own career. Although I had ensured that my animals were "cared for" in accordance with existing standards, I created for them an existence in which their attempts to express their basic natures were more or less thwarted at every turn. They led stunted, unfulfilled lives of boredom punctuated by episodes of fear and pain. Or worse.

The research I conducted for my PhD dissertation illustrates how I had used animal lives. I studied the learning abilities of two groups of adult monkeys who had been raised in entirely different ways for the first nine months of life. Members of the first group had grown up in laboratory facilities where they had access to their mothers and the offspring of other mothers inside a wire mesh pen the size of a small office. While not an unlimited free-ranging environment, the facilities permitted the monkeys to learn about being monkeys. They played, groomed, fought, sought reassurance, and practiced the acrobatic moves of young developing monkeys. At the end of the nine-month period they were housed in individual wire mesh cages where they could see, hear, and interact at a distance with other monkeys and the human laboratory staff.

The monkeys in the second group had been removed from their

mothers within the first day or two of life. That is, laboratory staff wearing thick, bite-proof catching gloves had seized them, frightened, shrieking, and grimacing, and pulled them from the arms of their mothers, who, with deep guttural barks and chattering teeth, clutched them desperately, struggling as if lives were at stake. They were then placed in individual cages inside large, thick-walled sound-attenuating isolation chambers. In these environments they could not see, hear, or even interact at a distance with other monkeys. These isolated monkeys experienced environmental changes only through their own behavior and the maintenance activities of staff technicians. After nine months these animals—who we referred to as "social isolates"—were removed from their confinement and placed in open wire cages like the members of the previous group.

From these two groups of monkeys I had drawn my experimental subjects. The data resulting from my tests demonstrated that the learning ability of the social isolates had been severely compromised by the social deprivation they had experienced during the most developmentally crucial portion of their lives. Although I could not see it in this way at the time, my observations of the social isolates suggested what any emotionally sensitive human child might have surmised to begin with: depriving young monkeys of social interaction and maternal love so thoroughly warps them that they cannot function normally and must suffer unimaginable anxiety, fear, and depression.

The crux of the story that I tell in this book is that I slowly became conscious of the animals' point of view and recognized that much of what I was doing as a scientist did not square with my own moral standards. This transformation had several different aspects—among them the realization that animals could in fact have a point of view and could experience pain and suffering as real as anything humans experience. Eventually I abandoned animal research and reeducated myself as a bioethicist. From this position I have spent many years reviewing proposed experimental protocols and participating in the oversight infrastructure created to regulate the use of animal subjects in research. Having taken part in both animal research and its ethical screening, I believe I am in a position to provide a balanced view of the use of animals in research, one that takes into account the legitimate needs of science and the ethical questions raised when animals' lives are appropriated for human ends.

In telling my story of ethical awakening I have several goals in mind. In addition to making amends for the significant suffering I caused Donna and other monkeys, I want to show how easy it is for scientists-in-training to fall under the influence of institutions that systematically set aside ethical considerations and often put laboratory animals into the same category as glassware and latex gloves. Fortunately, both students and established researchers using live animals in their work can counteract these forces and develop research approaches that make ethical issues as central as hypothesis development, experimental design, and data collection.

I offer my story in hopes that it will cause some members of the animal research community to take seriously the notion that research on animals should always present ethical questions. Unlike the all-or-nothing criticism often delivered by pained protectionists, the book develops the relatively moderate stance that consuming nonconsenting lives should, at least, occasion a strong sense of ethical ambivalence. I hope, too, that my discussion of research review by institutional animal care and use committees (IACUCs) will shed light on how to improve that process so that researchers might more meaningfully engage the relevant ethical issues, rather than just participating in a detached compliance exercise. The book is also written for the concerned public in that it provides a window into the otherwise nontransparent world of animal research and oversight. Finally, I hope that what is described in this book will help protectionists better understand scientists' plight and thereby develop more effective and constructive ways of engaging with them.

In telling this story I have chosen to describe a great many of the day-to-day realities of working in and around academic animal research laboratories. I want the reader to experience this world with me. Toward this end, I have narrated many examples of encounters with researchers, animals, and review committees throughout the book. These examples are presented to allow the reader to see the institutional- and societal-level dynamics at work and to understand how, as an individual enmeshed in them, I responded. I hope that my experiences and shifts in thinking can be taken as somewhat representative, and that readers can easily imagine themselves in my place.

While I am convinced that I now see the animal research world and the ethical issues involved with greater clarity, others will surely disagree. Some will say that I have become deluded, overwhelmed by sentimentality and emotional sensitivity. There is certainly some truth to this criticism. There can be no doubt that I sometimes have strong negative reactions to research practices, proposals, or incidents that involve animals. This occurred recently, for example, when I reviewed a National Science Foundation–funded grant application written by a wildlife biologist (and acquired by me through a Freedom of Information Act request) that described a study designed to test the heat tolerance of various species of desert birds in environmental chambers. While in the chambers, the captured birds would be exposed to increasing temperatures to the point that they could no longer compensate for the heat and would begin to desperately seek escape. This process was to be repeated several times with each individual until the bird's limits could be accurately determined. The justification was that the data so derived would aid in predicting the losses of desert bird populations as global warming advances. However, the thought of capturing sentient animals from the wild, holding them in individual cages, and taking them repeatedly to the edge of collapse—and on occasion purposely to death—was so offensive to me that it immediately overwhelmed the need to consider whether the study could have offsetting scientific value.

For me, some harms are so extensive that they cannot be offset by hoped-for benefits and cry out for a regulatory upper limit on untreated pain. At other times, negative reactions emerge only after I have analyzed the stated purpose of a study, examined its assumptions and methods, and come to the conclusion that the experiment is critically confounded and its justificatory thinking flawed.

In any event, the charge of sentimentality assumes that emotional and feeling reactions are always irrelevant to the ethical domain, which requires the application of only pure reason. But while raw feelings may not be ethically definitive in and of themselves, these reactions are at the very least instructive about the person's unstudied intuitions. To suggest that feelings do not have a place in the moral life is, I think, ridiculous.

Other criticisms will suggest that I fail to understand the requirements of science and scientific progress. Or perhaps that I have become a Luddite, a person at war with scientific and technological advancement. Maybe I long for an earlier world devoid of machines and desacralizing technology or a biblical paradise where "the wolf also shall dwell with the lamb" (Isaiah 11:6). This criticism has some merit. Although I have a generally favorable view of the value of medical research and technology and the benefits they bring, my view has been tempered many times by what I have experienced as an ethicist working on hospital ethics committees. I have seen many dying patients experience a great deal of unnecessary additional suffering when well-meaning but burdensome technical interventions were undertaken primarily because they were available, or because the attending physician was determined not to be "defeated" by death, or because the physician was invested in the outcome of a clinical trial. I have listened to the pleas of many patients, caught in this war against death, to be just allowed to die, or even to have their death hastened. I have also seen some of the vast injustice with respect to access to biomedical interventions—often those who are poor have no access to life-saving medications or surgery. My view of the benefits of biomedical and behavioral research has thus been humbled and contextualized—but I am no Luddite.

Some might argue that I was never truly committed to research in the first place and that by becoming an ethicist I have found a way to justify my modest productivity. In response I can say that I have not let this "cognitive dissonance" possibility go unexplored. As a clinical psychologist I have great respect for the notion that behavior is motivated by many factors, some not easily accessible to conscious awareness. Even so, I reject this possibility. It was the treatment of animals that I experienced and the ethical questions it raised that moved me away from the research domain, and not some *sub rosa* dynamic like fear of failure, fear of success, boredom, laziness, or hostility toward authority. As the saying goes in psychodynamic circles, "sometimes a cigar is just a cigar."

I have worried that the project of writing this book might appear narcissistic and self-aggrandizing. I feared that I would be seen as brazenly suggesting that my life and experiences were something special and that I was presenting myself as a model of virtue. The opposite is the case. Many other scientists have managed to broaden their perspective on the ethical challenges involved in converting innocent animal life into scientific data and have exhibited more courage and virtue than I. I have had conversations with dozens of ambivalent researchers, some of whom did not write about their concerns but quietly and courageously walked away from experimental animal use. Further, whatever moral development I may have achieved I owe to the prodding, encouragement, and mentorship of others, many of whom I came to know primarily by the luck of circumstance.

In this regard I owe a great debt to eight of the lab-animal veterinarians with whom I worked during my research career. These men, seeing their professional responsibilities as broader than just keeping research going like good soldiers, challenged me as they challenged themselves to improve the lives of animals in labs. My long intense talks and animal treatment encounters with S. Bret Snyder, DVM, changed forever how I saw animals and the predicaments they face in the laboratory environment. This education was continued by Michael Richard, DVM; Jerry Whorton, DVM; Daniel Theele, DVM; and Philip Day, DVM, at the University of New Mexico; Dan Hauser, DVM, of the University of Wisconsin, Madison; David Morton, PhD, MRCVS, of the United Kingdom; and James Mahoney, DVM, of the Laboratory of Experimental Medicine and Surgery in Primates at New York University. In similar ways, the campus animal-care staff members with whom I worked, Gilbert Borunda, Ector Estrada, and Roy Ricci, shared through example their empathic approach to research animal care and, equally important, never once lost their focus on maintaining the welfare of the animals in our facility. Their understanding of monkey needs and fears helped to ease the lives entrusted to them.

Many students' lives intersected with mine and left me the greater beneficiary. Numerous undergraduate and graduate students asked questions of the kind I avoided as a student, implanting tiny doubts that grew to become deeper challenges. Jeffrey Sproul, Robert Frank, Timothy Strongin, Harold Pearce, Alan Beauchamp, Jennifer Eldridge, Terri Moyers, Janet Brody, Lee Davis, William Kuipers, Martha Carmody, and Beth Dettmer stand out as particularly influential. They were all strong-willed skeptics, not intimidated by the expected roles of teacher and student.

Early in my developing ambivalence about animal use, the philosopher Bernard Rollin acknowledged my ethical plight and assured me that I was on the path to discovery and not ruin; for this I am forever grateful. Ken Shapiro, founder of the Animals and Society Institute, illustrated to me the scholarly place of animal welfare issues; my continuing collaboration and friendship with Hope R. Ferdowsian has been a regular source of important perspective and direction. Jon Lewis, a fellow Harry Harlow mentee and friend, continues to provide strength and outspoken reflection. The many important discussions I have had with Gay Bradshaw of the Kerulos Center, Byron Jones, Diann Uustal, William Gannon, and Alexis Kaminsky about ethics education in science have left me hopeful about the eventual success of this endeavor. I cannot overestimate the importance of my long-term collaboration with the physician and bioethicist Cynthia Geppert. She constantly offered her multiple competencies to me when I was in need of intellectual and emotional traction. Her recommendation that I seek expert and caring help from Heather Wood and Lane Leckman in negotiating the psychological dynamics of this project is only one example.

I could not have become the author of this book were it not for the academic privilege of sabbatical leaves. This privilege permitted me to pursue two fellowships, one in clinical psychology at the University of Washington Department of Psychiatry and Behavioral Sciences, and one in bioethics at the Kennedy Institute of Ethics at Georgetown University and the National Institutes of Health. At the Kennedy Institute I had the incredible good fortune of working with philosopher Tom L. Beauchamp and fellow scientist Flora Barbara

Orlans. Tom and Barbara are responsible for pushing me to the point at which I could contemplate writing this book. I could not have had better mentors and friends. I deeply regret that Barbara is not alive to see the completion of the book.

My series editors at the University of Chicago Press—Jane Desmond, Barbara King, Kim Marra—along with acquiring editor Douglas Mitchell, supported me in so many ways, in part by insisting on rigorous and careful prepublication review. Jane answered all my feverish e-mails within minutes and always with sound advice. My developmental editor, Eric Engles, and manuscript editor, Ruth Goring, transformed and improved my manuscript in ways that I could not have imagined possible at the start of our work together.

As a final note, I want the reader to know that writing this book was a very difficult process for me. It involved reliving parts of my life during which my character and moral courage were not up to the ethical challenges that confronted me at the time. In addition, some of the narrative of change risks offending people who supported me throughout my education and life as an independent researcher. Jim Sackett and Helen Leroy are particularly important in this regard. I do not want to lose the respect of these people, but I recognize that this might be a cost of publishing this book. In the end I decided that I had to go ahead regardless.

On the positive side, this project put me back in touch with models of kindness, wisdom, and humility that were present all through my life. My parents, John and Dorothy, were models of how the moral life should be lived day to day. My mother's determination and my father's quietly heroic life as a New York City firefighter regularly illustrated the place of courage in the development of character. Luckily, most mornings of my life as a professor at the University of New Mexico began with a talk with my colleague Peder J. Johnson. In those conversations we helped each other tame the avarice of achievement by placing our work centrally in the broad context of our lives. My son Jeremy's effortless kindness to the animals in our house and my young daughter Katie's ethically motivated vegetarianism and regular animal rescues each illumined to me such natural concern for innocents that I could not finally miss the lesson. Most importantly, my wife Charlene McIver brought her quick-study brilliance, warmth, and support to bear so many times as I worked through the issues described here that she could just as well be the author. Many of these people, and others not mentioned, presented me with challenges that stood like gifts to be accepted or ignored. Unfortunately, my appreciation of many of these gifts was very much delayed.

Introduction

Throughout human history we have had a tremendously difficult time understanding where animals should fit on the continuum of our ethical concerns. To grasp the depth of this confusion one has only to consider some of the different roles that animals have played and continue to play in the human imagination. Animals have been understood to be gods, models of strength and virtue, possessors of eternal wisdom, healers, entertainers, and companions, but they are also seen as threats to life, embodiments of evil, adversaries, and vermin. As sources of food, animals are given a status that can range from sacred being to mere nutrient factory. Many people avoid any food or product that requires killing an animal; others consume meat without giving a moment's thought to how animals are treated in the industrial food system from which it comes. In our homes, some animals are given the kind of affection we usually reserve for children and grandchildren, while others are despised as harmful invaders. The upshot of this ethical confusion is that animals in general are treated with an inconsistency that is spectacular in its breadth. The lives of animals can matter not a whit, or they can matter very deeply. The determining factor is often the meaning that the animal in question has for humans.

In biomedicine and in my own field of psychology, ethical questions about using animals in research have long been fraught with controversy because they arise from sharp contrasts. On one hand, using animals as experimental subjects often involves undeniable harm to the animals: their bodies are sliced, their organs removed; they are shocked, drugged, exercised to exhaustion, starved, confined in tight spaces, and sometimes deliberately killed. On the other hand, the procedures causing these harms are designed to advance scientific knowledge, and the knowledge so derived can become the foundation for development of drugs and therapies intended to cure human disease and alleviate human suffering.

Thus the basic ethical questions in biomedicine, psychology, and other fields using animals are these: Do the potential benefits of a research project justify the harm inflicted on animals? Should there be an upper limit on the amount of pain that is acceptable for researchers to create, regardless of arguments of a project's potential usefulness? Do animals possess rights of noninterference that restrict their use by humans? When harms to experimental animals are taken seriously and considered thoroughly, these questions should pose complex dilemmas.

In 1881, at the seventh World Medical Congress in London, Frederick Goltz and Daniel Ferrier debated whether the parts of the brain functioned together as a whole—the equipotentiality theory—or whether separately located components had clearly different functions. Goltz demonstrated his evidence for equipotentiality by presenting to the audience a live dog that had had large sections of its brain removed; he showed that the dog appeared to behave normally as it moved around the stage. In Goltz's mind, this demonstrated how the remaining brain tissue was quite capable of carrying out all the necessary perceptual and control functions.

Then, to support the separate component position, Ferrier presented a macaque monkey who also had sustained experimental brain lesions. Its behavior was very different from the dog's: the monkey limped and dragged its body across the stage, appearing to have lost motor control on one side. Upon seeing the disabled monkey, the prominent neurologist Jean Charcot spoke out from his seat, declaring in French, "It is a patient!"

Charcot's statement, whether meaning to or not, captured an essential part of the ethical dilemma of animal research. He pointed out that while the monkey's appearance showed that it might be useful as a model of a human patient's painful neurological pathology, it was a patient itself—that is, it too suffered and was obviously in need of care and treatment.[1] In this way Charcot highlighted the conflict that faces the experimental scientist: either acknowledge the reality of animal harm and work out its ethical relevance, which might require limitations, including the possibility of not proceeding at all; or instead deny, ignore, or minimize the animal pain and suffering behind a veil of uplifting real or only hoped-for human benefits.

The question residing in the deep structure of Charcot's declaration was whether humans are the only ones that matter ethically in the experimenter–animal relationship. Is the connection a relationship at all, or is it just a matter of owners using their acquired property for their own purposes?

Unfortunately, throughout history and even to this day, many scientists and researchers have chosen to ignore animal pain and suffering and treat animals more or less like owned objects. They have stacked the deck in favor of deciding for animal use by minimizing the costs involved in the cost–benefit comparison or denying that the costs are real or even of ethical consequence. This has resulted in an unbalanced concern for human over animal interests, a valuing of hypothetical or potential human benefit over real animal suffering.

There are many notable examples in medical history of manipulating animal life for human benefit. In 1964 James Hardy transplanted a chimpanzee heart into Boyd Rush, a sixty-eight-year-old cardiac patient deep into the process of dying, who continued to survive in a drugged and unconscious state for a total of ninety minutes following the surgery.[2] It is illustrative of the ethical thinking of the time that in Hardy's report of this milestone event not a single word is written about the costs to the chimpanzee, probably wild-born, in terms of anxiety and pain and its loss of a natural chimpanzee existence in order to provide those additional ninety minutes to a human being who was about to die—and we can't even be sure that these were additional minutes or whether the patient would have survived longer without the surgery. In fact, the only detail about the animal that is mentioned was that its heart pumped 4.25 liters of blood per minute. Not even the animal's sex is indicated. In the end, the chimp was transformed by the research context of the day and a highly respected surgeon into nothing more than a container and source of life support for an organ. One need not look very closely at such episodes to see the desperateness that can underlie attempts to improve and extend human life and the one-sided emphasis regarding whose suffering matters.

St. Augustine observed in the fourth century that human curiosity and the drive to know—especially when disease and the fear of death are involved—have an essential carnal quality, generating a strong desire that can escape the bounds of reason. The strength of this drive, which St. Augustine called *libido sciendi*,[3] is given expression in many core elements of Western culture, such as in the myth of Prometheus, in which Prometheus steals fire from Zeus in order to improve the human condition. Tellingly, however, in the same myth we are also reminded of the dangers inherent in a single-minded search for knowledge, as Prometheus's exploits cause the release of the contents of Pandora's box and his own eternal torture. Similar warnings abound: in Francis Bacon's influential book on the new experimental method, the *Novum Organon*, published in 1620, he enjoined readers to apply the experimental method with care and reflection because it could lead to abuse of nature; in *Frankenstein*, Mary Shelley illustrated the tragedy associated with pursing progress at all costs; and J. Robert Oppenheimer confessed that scientists had "sinned" during the Manhattan Project.[4]

In the *New England Journal of Medicine* in 1966, Henry Beecher described twenty-two obviously unethical human experiments, including one in which the investigators injected cancerous melanoma cells harvested from a child into her unsuspecting mother in order to study her immune reaction.[5] He was careful to implicate as the causal factors of these transgressions not the responsible individuals' bad character but career pressure and the strong motive to reduce suffering and the threat of disease. In this powerful exposé, Beecher chose to protect the identity of the researchers by not providing the exact references of the work discussed. The cases he described illustrate clearly that in the passion for medical advance, even the moral status of humans can be ignored.

To advance knowledge in biomedicine and psychology, researchers generally use animals for a very particular reason. Mindful that causing harm to human subjects is patently unethical—Beecher's examples notwithstanding—they do to experimental animals what it is forbidden to do to humans. Scientifically, this is justified because the mammals used in most research—rats, dogs, rhesus monkeys, chimpanzees, and others—are so biologically similar to human beings. In the case of primates like monkeys and chimps, the similarity is in fact extremely close. Humans and chimpanzees share 98.8 percent of their DNA. Thus, depending on what is being studied, results obtained in animal experiments can at times be validly extrapolated to the human species, with appropriate qualifications. In this way, experimental animals serve as models for humans.

Although the practice of using animals as models for humans when direct experimentation on humans would be unethical is firmly entrenched in biomedicine and psychology, it rests on an almost hypocritical inconsistency. When researchers want to extrapolate their animal results to humans, the close biological similarities between animal subjects and humans are celebrated and emphasized. But when it comes to justifying doing to animals what can't be ethically done to humans, any similarities between animals and humans that might have ethical relevance—the experiences of pain, fear, and distress, the ability to suffer—are minimized, ignored, or denied.

To avoid seeing this double-standard treatment as a problem, and thus to avoid ethical reflection on it, requires an extraordinary level of self-deception, rationalization, and selective blindness. It requires us to objectify experimental animals, categorically excluding them from the class of beings requiring full ethical consideration. It requires us to evade manifestations of harm like pain, fear, anxiety, depression, grief, and anguish in animal subjects, and and to convince ourselves that animals can't possibly feel what humans feel.

I can make these claims about what goes on in the minds of many animal researchers because I was one myself for the bulk of my career. I conducted my graduate school research under the direction of Harry F. Harlow, John W. Davenport, and Gene P. Sackett at the Department of Psychology Primate Laboratory associated with the Wisconsin Regional Primate Research Center[6] and the University of Wisconsin, Madison. When I first arrived as a graduate student in 1968, Harlow was already a psychology leader and celebrity because of his studies of attachment and mother love in monkeys, and soon—late in his career—he was to become publicly infamous in animal protection circles for conducting what has been referred to by some as cruel and "useless" experiments with monkeys.[7] For my dissertation research I compared the ways in which monkeys with extensive social experience and monkeys that had been raised in isolation learned to adjust to different patterns of reinforcement in a simple situation where they had to press a response lever to get access to food rewards.

After completing all the requirements for a PhD in psychology at the University of Wisconsin, I was hired as an assistant professor at the University of New Mexico, where I began teaching and research in 1971. Hired with the clear expectation that I would start a primate research laboratory, I devoted my first few years at New Mexico to doing just that.

In the early days of this process, while my laboratory was taking shape, there was a thrill that accompanied going to the lab early in the morning to get things started. I could hear the excited sounds of the monkeys, then housed in rows of individual cages, as I opened the doors to the animal colony. I could see them rattling their cage doors, turning somersaults repeatedly in the small area of each cage. I was happy to see them, and they were not anonymous experimental subjects: I recognized each of them individually.

The first task was to feed them; I poured commercial monkey chow into the feed cups attached to their cages and filled their water bottles. I watched with satisfaction as they grabbed the biscuits, ravenously stuffing them into their cheek pouches. Their facial expressions shifted rapidly as their glances skittered between me, the other animals, and the food. Some monkeys made submissive lip-smacking gestures, while the aggressive ones tried to engage me eye to eye while bobbing their heads and making deep guttural sounds, looking as though they would attack me if the opportunity presented itself.

These encounters, though raucous and tense, brought me a sense of deep satisfaction. The animals were my ticket to discovery and academic promotion.

I managed to construct the belief that the monkeys and I were in the science enterprise together, despite some clear signs that the monkeys saw me as an enemy, or a least a fear-provoking object. I thought of them as partners in research. I was invested in them and they in me. It was as though we had a mutual understanding that benefited both of us. I believed that as long as I provided them with nutritious food, clean air and water, safe cages devoid of sharp edges, and excellent medical care, I was discharging my ethical duties to them. The absence of obvious disease was the animal welfare standard of the time.

The only slight crack in the facade suggesting that I understood something about the monkeys' real predicament was when I described them, as I often did, as "draftees in a just war against human suffering." I used that language knowing full well from my own conflict with military service and the Vietnam War that a draftee was a person typically torn away, without consent, from his own life plans and preferences in order to participate in a task, often dangerous, that was judged by others to be a worthy use of his life. There is a huge ethical difference between a partner and a draftee, but the distinction was lost to my conscious sensibilities.

I insisted for many years on believing the comforting story that the monkeys were my partners in the search for knowledge. However, slowly I began to allow myself to see that the monkeys' morning "greetings" were mostly displays of desperation, frustration, and anger brought on by the poverty of their living situation. The fact of the matter was that these socially and cognitively complex animals were locked in boxes twenty-four hours a day, seven days a week, except for the time they were transported to participate in various experiments that lasted an hour or so. And the experiments I conducted were done in experimental chambers (i.e., boxes) even smaller than the ones they lived in.

No matter how clean the cages were kept, no matter how many fresh air exchanges took place each hour in the rooms, no matter how nutritious their food was, and no matter the perceived or real importance of the experiments, nothing could change this reality. Nothing could change the bare fact that this laboratory, like so many others, systematically deprived the monkeys of the possibility of living a form of life for which their evolutionary heritage had prepared them—a form of life devoted to touching, developing relationships, exploring novel environments, climbing, foraging for food, being hungry and frightened, having sex, raising infants, fleeing predators, and fighting. The laboratory environment permitted only pitiful fragments of some of these potentials. The monkeys engaged other animals in the colony rooms, but at a distance. They continued to threaten, present their bodies to be groomed by individuals hopelessly out of reach, and otherwise keep track of the other monkeys in the other cages the best they could. I came to realize that I was the only member of the "team" who was really benefiting from this relationship.

The justificatory argument that I accepted at that time was that these crushing limitations and any resulting boredom, pain, or distress were made worthwhile by the basic and applied knowledge that was gained from the research. My sensibilities to the actual consequences of the research restrictions were dulled by my casual acceptance of the unanalyzed truth of this justificatory argument. I avoided the reality of the nature and impact of my professional activities in a variety of subtle ways. For example, I carefully sanitized the language I used to describe the research. The monkeys were "housed," not confined or held in empty stainless-steel mesh boxes that restricted their movement and offered little opportunity for the monkeys to do what monkeys naturally do. Newborn monkeys selected to be raised in isolation were "removed" from their mothers, not torn from their arms shrieking and terrified. It seems so obvious to me now that if you must sanitize the language that is used to describe the procedures in regular use, you have entered morally perilous territory and the situation calls out for reflection, reconsideration, and evaluation.

The thirty-plus monkeys that participated in my early research are all dead now but remain in my memory as a group of ID numbers and names attached to images of their distinctive attributes. E-1 was tall, thin, and constantly frightened. E-7 had an Asian look to her eyes and a mass of scar tissue on the top of her head because she continually scraped it when she was transported from her "home" cage to testing situations. I named E-38 Fuzzy because she had a completely furry face like a small cat. G-49, or Moose, got her name because she was large, slow moving, socially confident, nonaggressive, and friendly to the monkeys she encountered during social ability tests. E-15 was slender with odd repetitive behaviors. The reddish-colored E-25 liked to line up her monkey biscuits in an absolutely straight line on the floor of her cage before she consumed them. J-90 was a bully who chased and fought with subordinates much smaller than he was. I could go on.

What I have learned since then is that while I knew these animals well from spending many hundreds of hours watching them, for the most part I did not really see them. What I mean is that at that point in my career I did not comprehend their predicament. I did not appreciate the extent of the monkeys' vulnerability and potential for suffering. I saw only my own scientific needs and the active approval of my colleagues.

As I slowly began to appreciate their circumstances, however, I began to see them as whole beings, and questions about the ethical justification of animal research began to dominate my thinking. What had been an unquestioned mission became full of ambivalence and uncertainty. Glib arguments about the importance of research began to feel thin and rote. I could no longer believe that the co-opting of other animal life for our benefit was obviously just. I began to appreciate the extent to which I was never really encouraged by my mentors (and my own character and conscience) to seriously engage the central ethical question: how do the real or potential benefits to humans justify the costs to other sentient species, who would strenuously avoid these experiences if they were not actively prevented from doing so?

For me, what was required to begin to open the door to serious questions about ethical justification was making a shift from (1) just being impressed by the cognitive and emotional capabilities of the animals with which I worked, or which I read about, to (2) seeing that these characteristics had ethical relevance because they indicated that primates are very much like humans emotionally and in their capacity to suffer. I think now about how often in my lectures on animal learning and cognition I would emphasize studies like the language work of Ann and David Premack and the chimpanzee Sarah. In their research the Premacks used pieces of magnet-backed irregularly shaped plastic to represent words and tested the chimp's ability to answer questions and write communicative sentences. At one point they asked themselves whether the plastic symbols had really become words. That is, would Sarah describe the plastic symbol in the same way she would describe the real object if it were present? So they presented the symbol for apple, which was a blue triangle, and asked Sarah to do a feature analysis. They asked her via their system whether the symbol for the word *apple* was red or green. She answered red. Next they asked whether the symbol was round or square. She responded round. Did it have a stem or not? She chose the symbol for stem. Was the plastic symbol a square with a stem or round? She chose round, indicating that for her, the roundness was a more basic characteristic of an apple than the stem, which only occasionally was present on actual apples.

Each time I used this and other examples, even during my early animal-use career, I felt an intense emotional involvement that went along with describing the phenomenon and went beyond just intellectual excitement. It was as though describing this kind of human-like cognitive complexity was touching something very important, but I did not dig into what that was. I could imagine Sarah sitting across from Ann Premack, both hunched over the symbol board where the two of them were trying to dissolve the species barrier and each describe to the other how they saw and understood the world. The Premacks were trying to get a grasp on the fundamental nature of language and what was unique about human use of it, and Sarah's responses were helping to clarify that distinction.

In the early days I saw the reaction as validating my interest in animal research—showing what complex creatures nonhuman primates are, and by extension how important I was for working with them. However, as the years passed and my sensitivity to animals like Sarah and her predicament sharpened, I saw a chimp cooperating with a scientist not as a partner but simply to get along in a human world that was so confining compared to the one from which she was captured as a youth. Now there was a clear emotional chill. It was about the fantastic lack of balance between scientific advance and animal costs. The result was that I no longer felt so proud about being a monkey-life thief, even though I was on an interesting scientific quest.

I am describing my transformation—from an avid animal researcher to someone who doubts whether most present research on animals can be justified—as a process of personal development and expanding awareness, but in truth it had many contextual components. Challenges to my narrow way of thinking were offered by other individuals, circumstances, and, of course, the animals themselves. In the chapters that follow I try to highlight these external forces.

Further, I did not enter the animal research arena already blind to animal suffering and convinced that research on animals was always justified because of its potential for human benefit. The frame of mind needed to experiment on animals without twinges of conscience was actively inculcated by my peers and mentors and by the practices, language, and unwritten rules of the scientific enterprise. That initial process of "indoctrination" is a central topic of the first chapter.

Beyond the realm of ethical concerns, from the vantage point of my new perspective I have been able to see more clearly that the approach to animal research that assumes animals to be nonfeeling objects is self-limiting. The refusal to see the whole animal, to admit the validity of emotion and subjectivity in both researcher and animal, makes it impossible to understand animals as animals. Deprived of this knowledge, researchers acquire data that is at best partial and lacks the context for proper interpretation. For this reason, integrating ethical concerns into problem identification and research design—in the thoroughgoing way demanded by the view that animals are sentient beings with ethical standing—may not only limit animal suffering but also yield better science.

1
Erosion

Be sure you do not let anything, not even your
good kind heart, spoil your experiment.

MAX GOTTLIEB TO MARTIN ARROWSMITH IN
SINCLAIR LEWIS'S NOVEL *ARROWSMITH*

Most people, I think, do not reach early adulthood with a frame of
mind and set of values compatible with experimenting on animals.
Growing up — especially in a household with pets — children learn
to appreciate animals as beings with inner lives much like themselves
simply by applying their naturally developing empathy and intuition.
Any tendency for society, with its sometimes ruthless objectification
of animals, to refute this working assumption is counterbalanced by
the teaching of the general "do unto others" ethical framework and
a maturing sense of compassion for those who suffer.

The typical college-age person is therefore likely to view with
some concern and twinge of conscience the implantation of electrodes in rat brains, delivery of painful electrical shocks to monkeys,
and administration of powerful drugs to dogs. For this reason, those
attracted to a career involving research on animals must undergo
an emotional and ethical retraining process every bit as important
as their scientific training. They must learn to put aside identification with animal pain and suffering and replace it with a passion
for advancing scientific knowledge. They must come to believe that

they are joining the ranks of a special corps of truth-seekers and improvers of the human condition who may assume their work to be justified on its face, no matter the cost in animal lives.

Such was the case for me. I grew up with deep emotional attachments to family pets, believed without question that animals had internal lives that mattered to them and were capable of feeling joy, sadness, fear, disappointment, and pain, and was revolted by cruelty to animals. By the end of high school, I had a well-developed sensitivity to suffering and a good sense of compassion. I also had an abiding interest in reducing the human suffering that I saw in the world, and when this interest found its vehicle in experimental psychology, I took a path that brought my career ambitions into conflict with my natural inclination to abhor the deliberate harming of animals. By the time I had finished my undergraduate education and started graduate school, my professors—and the overall research context into which I threw myself—had exorcised my sentimental concern for animals' welfare and constructed for me a new belief system in which there was really no such thing as the animals' perspective.

My father John was, not surprisingly, very influential in modeling how I should comport myself and treat others. A New York City firefighter and part-time longshoreman during most of my youth, he projected a quiet strength and protectiveness. When he returned from a full day of work, his clothing had a distinctive smell, a combination of smoke and bananas, which became for me the perfume of security. While there were many hints of a rough-and-tumble life at the firehouse and on the docks, he was a gentle man at home. He loved children—and not only those in his own family. When he and I attended local athletic events together, he would often engage the small children who sat near us, talking and laughing and giving them pennies. He also loved dogs. Although his expression of affection for the animals was muted, all of the dogs we lived with while I was growing up waited at the door for his return from work and then rested near him wherever he sat.

When Dad's schedule allowed him to be home on the weekends, he and I spent countless hours playing catch with whatever ball was in season and running around an undulating cinder track in a WPA-constructed park called Victory Field. There we watched local baseball, softball, and semi-pro football games until we were quite familiar with many of the players. Not much was said between us during those times; rather, they were dominated by the relaxed feeling of being together without a particular goal that needed to be accomplished. I knew my father wanted me to develop into a strong person with streetwise sense, as insurance against the risks he knew were out there waiting. If there was a demand that came from him, that was it.

My mother Dorothy was a cautious person, skeptical about the goodness of the world. Like my father she was experienced with the ins and outs of city life and actively counseled defensive vigilance. Significantly for me, however, she spoke her mind with confidence and strength, and she both valued and exemplified independence. She expressed affection more directly than my father, but firmly believed that love had its limits when it came to her personal and family relationships — except, that is, in the case of the family dogs. For them she was forgiving and effusively kind. She refused to buy commercial dog food, preferring to cook the dogs' meals of meat, rice, and vegetables. My mother had been encouraged by her parents to pursue higher education, but for her own reasons she left an academically oriented high school in New York (Julia Richmond) for a course in secretarial skills. She ended up working — contentedly, I think — as a secretary to the principal of an elementary school in Queens. She prided herself for her meticulous memory and organizational skill and believed she would have made a good doctor. I think she was right. Though she would have been happy for her children to have solid jobs in any respectable venue, she encouraged academic interests like her own mother had done for her.

My family occupied a comfortable two-room basement apartment in the house owned by my grandparents. My parents and sister slept in two of the three upstairs bedrooms, and I had the foldout sofa bed to myself in the living room of our apartment. While I did not have my own room, the arrangement did provide me the privacy to stay up and watch the *Late Show* and listen to the *Symphony Sid* jazz radio show late into the weekend nights. Most nights I could also invite the family dog to share the bed with me, in violation of my mother's rules. The dog and I were typically awakened in the morning by the sound of footsteps on the stairs from the upper floor, giving the dog plenty of time to jump to the floor and take up a more acceptable location.

In the immediate neighborhood where I grew up were three very large cemeteries. The cemeteries provided places for my friends and me to hang out in the late afternoon and evening, away from the prying eyes of neighbors and police patrol cars. Many of the grave sites had gray cast-iron furniture, the original purpose of which was to facilitate long visits by the friends and family of the deceased. We used the tables and chairs to just sit around, play cards by flashlight, talk, and, as teenagers, drink beer together. We meant no disrespect when we shared our drinks with the deceased by toasting them and pouring a little on the grave. It was a way to thank them for the use of the furniture and the much-valued privacy.

The cemeteries were also homes for the local wildlife. They were full of squirrels, many different kinds of birds, rabbits, and some feral cats. The squirrels were tame, readily accepting peanuts and eating leftover sandwich debris and potato chips. Many of the cats still wore collars from previous human associations and were sad to behold. Some would approach us seeking a kind touch; others crouched tightly on the periphery, too frightened to approach yet appearing to be trying to overcome their resistance. One day I saw a very large bird, about eighteen inches from beak to tail, with a deep, full chest and feathers of an orange-rust color with black highlights, perched proudly on a Civil War soldier's headstone. None of us who saw the bird had any idea what it was. Roy Johnson, the most intellectual of our group, later searched his encyclopedia and identified the bird as a pheasant. From that point on, birdwatching was added to our daytime cemetery activities.

Local shopkeepers and neighborhood notables, each with a colorful past and defining life story, provided me with abundant examples of the variety of personality and approaches to life that can be found in human society. A few of these people I remember with particular clarity. Abe the grocer, for example, was a short, pugnacious man with a crude, faded tattoo made up of a series of symbols and numbers on the inner portion of his left forearm. I learned from my father that the tattoo had been applied while he was in prison during World War II. Later, when I worked for him delivering groceries and stocking shelves, he told me stories about his time in Auschwitz. He was the first adult man I ever saw cry openly.

Based on what I saw in my family and neighborhood, I grew up believing that animals had a special place in human lives. For example, while my grandmother frequently fought bitterly with my grandfather and acted ambivalently toward my sister and me, she lavished affection on her dog Buddy, a beautiful white spitz. When I was about four years old, I tried to pull a small gardening tool away from Buddy, and he took exception and grabbed and held my right wrist in his mouth. My mother heard my cries, ran out, and intervened. Buddy's teeth had left a bleeding puncture wound, so she quickly took me to the doctor's office on the next block; he applied a few drops of acid to clean the wound. When we returned home, the facts of the incident were investigated with a neighbor, and in the end it was determined that I was primarily at fault. I was punished, while Buddy was compensated for his premature scolding with a special dinner.

My family had two dogs while I was growing up: Prince, an oil-black cocker spaniel, and then Penny, a medium-sized black-and-white mixed breed that my father bought from a cab driver for two dollars. My sister and I took turns walking each dog and caring for its needs. Each of them brought warmth and joy to our lives on a daily basis and was clearly a member of the family. For some unknown reason, however, neither Prince nor Penny lived very long. Their deaths were greeted with open, unembarrassed grief from the entire family.

When Penny died, the pain and sadness ran so deep that my parents decided that we would not get another dog. Instead my mother purchased a bright green and yellow parakeet that she named Paddy. She believed that the bird, while welcome in the house, would not elicit the same amount of attachment and subsequent pain when he died. Several times a week after dinner, my mother would hang an old bedsheet in the entryway between the kitchen and the living room. Then Carol and I would take turns opening Paddy's cage door so that he could stretch his wings. Paddy flew joyfully around the room, landing on lamps and tables and people's heads. It was a ritual that we all enjoyed, but it was clear that it was done for Paddy, an acknowledgment that he was spending most of his life in a cage and that was not quite fair. When Paddy was found cold and still in the bottom of his cage some five years after joining our household, it marked the end of the family's pet ownership. There was no way to escape the grief and sadness. Intimacy always engenders a vulnerability to loss, no matter the attempts to stay emotionally separated.

I can vividly recall the ragman coming down the alley every week or so in a rickety wooden wagon pulled slowly by a brown horse. From anywhere in the house I could hear the clanging of the cowbell that was attached to a pole to the right of the wood-plank front seat, and the rhythmical *clip-clop* of the horse's hooves on the hard cement. My grandmother would take me out to meet the man and the horse. She gave me treats and taught me how to feed the horse as she spoke to the driver in German. I was intimidated by the large animal at first but quickly came to enjoy his strong smell and the feel of his mouth and warm breath. I could see that he liked getting his forehead scratched and his neck and shoulders rubbed. That I become comfortable performing these acts of simple kindness seemed very important to my grandmother. To my memory, this was one of the few explicit lessons she intentionally taught me. While I enjoyed these encounters, I also became aware of the plight of both the ragman and his horse, who spent their lives going through other people's garbage in order to survive. I could see the man's tattered clothes and thin, drawn face, the protruding hipbones and ribs of the horse. When I asked my grandmother to help me understand these issues, she would simply say, "Das leben ist hart"—life is hard. I slowly began to understand what she meant.

Several times in my childhood I witnessed or heard of acts of cruelty to animals. I know that these experiences left a deep impression on me because I remember them very clearly. One of them occurred in the early 1950s, when I was about ten years old, on a visit to the home of my aunt Evelyn and her family, who had moved to Hempstead, Long Island. The small, comfortable house they had purchased sat on an acre of land that adjoined a tract of woods that had a small stream running through it. My cousin Robert and I immediately took off for the woods, to spend the day exploring.

When we arrived at the stream, three or four boys a little older than ourselves were huddled together on the bank staring into an aluminum pot. Peering over their shoulders, I could see an eel in the bottom of the pan along with about an inch of murky water. The eel curved around the inside edge of the pan, and it slithered in circles as the boy holding the pot reached down to poke it. Robert, who knew the boy, quickly jumped back and yelled at him to leave the eel alone and throw it back into the water. The boy ignored him.

I didn't understand why Robert was shouting until I looked more closely and realized that the boy held a small pocketknife and was trying to stab the eel with the point of its blade. Whenever he moved the knife toward the eel, placing the tip of the blade on its skin, it slithered away before he could penetrate the eel's body.

Robert shouted at him again. This time the boy looked up angrily and said he really wanted to see what the eel looked like on the inside and Robert had better leave him alone. Some of the other boys chimed in and glared at us threateningly.

Robert and I came to an unspoken agreement that the eel's life was not worth the beating we were likely to get if we continued to intervene, and so we walked off. I was sorry for the eel and felt like a coward. From the look on Robert's face I could tell he was feeling the same. As we walked away, I remember wondering, what *did* the eel looked like inside? We returned to the house, where we got caught up in a play with Robert's sisters, and laughter soon replaced the feelings of fear and cowardice.

Another exposure to cruelty to animals took place when we visited the home of my father's sister Edna and her husband Russell in rural Connecticut. Off to one side of their rustic house was a large boulder into which my uncle had chiseled a cat's head. The date of its completion, 1944, was also engraved neatly into the base. He explained that the sculpture was a memorial to one of their beloved cats, Nosey, a mostly brown-and-gray cat with an odd dark marking below its nose that looked like a mustache. The story he told was that after the cat had gone missing for several days, its body was found nailed to a tree. An accompanying note stated that the cat had been tortured because its face reminded the perpetrator of Hitler. I was horrified by the story and frankly could not fathom the depth of hate that it represented. Why would someone who hated Hitler choose to express it by crucifying a cat?

In school, I was prone to distraction and generally did just enough work to get by. However, as a result of prodding by my family and the efforts of my father's younger brother Ed, a worldly and curious merchant seaman, to get me interested in reading, by the time I reached high school I was more academically motivated. I also studied the clarinet and saxophone and spent hours each evening practicing. I had great fun performing rock and roll and jazz with small local combos at dances and café venues. A group of my friends and I were attracted to the Greenwich Village bohemian scene, where we could listen to music and poetry readings for hours for the cost of a cup of coffee or hot spiced tea. These experiences encouraged more serious reading of authors like Sigmund Freud, Carl Jung, the existentialists Camus and Sartre, and the beat poets Lawrence Ferlinghetti and Allen Ginsberg. These explorations transformed me: I started to become a more mature, skeptical, and reflective thinker. I began to read books on biology, American history, and geology. Science began to take more and more of my time.

Nonetheless, by the time of graduation from high school I had not seriously applied to any colleges. I began to think that I might skip college and take a city job like my father or focus on music as a profession. In an attempt to keep my academic options open, however, I enrolled at the Bernard Baruch School of Business and Public Administration of the City College of New York (CCNY), taking classes in accounting, marketing, and economics. But I failed to connect with the business curriculum; I didn't want to learn about ways to manipulate the consumer and keep track of profits and losses. In protest and disgust I just walked away from the classes, allowing a string of incompletes to fill my academic record.

Still, honestly evaluating my musical skill and the likelihood of consistently making a living as a musician yielded the conclusion that while I was competent and even talented, I was not star quality by any stretch of the imagination. Facing these facts, and considering my father's advice to not follow in his footsteps, I began to think more seriously about going to an arts and sciences college.

During this time of turmoil, I was heavily influenced by the terrible turn my father's health had taken as I was growing up. It was a harsh irony that this man who made a living with his strength and quick wit had been beset with Parkinson's disease in his late forties. I was about eight when I had first noticed his right hand shaking from time to time. At first I did not recognize the tremors as a sign of disease. I recall asking him to *make* his hands shake, as if it were some kind skill. Eventually I had become aware of the shaking as a sign of a serious illness, and then acutely aware of the sense of hopelessness that it engendered.

As I pondered what direction to take in life, anxiety about the continued worsening of my father's symptoms was an inescapable part of my daily reality. His condition, and the mental and physical health problems experienced by other family members, had made me starkly aware of the cruelties of disease and old age. I was seeing up close the life-destroying repercussions of mental and central nervous system disease and the pitifully lame clinical interventions that were available to provide relief. I wanted badly to find a way for people to escape this condition of helplessness. At the same time, the words of family members began to resonate: opportunities to pursue educational advancement must not be squandered, because, like one's health, they may not always be there.

I finally decided to apply to colleges and study some area of science related to human health. For various reasons I wanted to leave the New York City area and, acting on the advice of the coach of the semiprofessional football team with which I was involved, chose to attend Texas Tech University in Lubbock. Once I was accepted, the registrar arranged for me to first attend Lubbock Christian College

(LCC), where I could add a few A's to my record and in just one semester erase all those incompletes I had received at CCNY. Then, if successful, I would be able to register at Texas Tech.

*   *   *

My semester at LCC was academically unremarkable—aside from earning the necessary A's—but the introduction it provided to life in Texas was transformative. Virtually every step I took during my first days in Lubbock presented me with a view of life that was unfamiliar. The way people dressed, the twang in their speech, the endemic friendliness, the openness of the arid terrain—it was all very new to me.

On my first day in town I asked the counterman at a café how much it rained in Lubbock. "About fifteen inches a year," he responded.

"That's all?" I said, truly surprised.

"I know that doesn't sound like much," he continued, "but you ought to be here the day it comes." We both laughed hard at what was certainly an old joke for him.

I absorbed all this novelty quite readily, finding that I rather enjoyed Texans' affability and easygoing nature. By the end of the semester, though still a "Yankee," as my friends put it, I had settled in.

One aspect of Texas culture that had great impact on me was the way people related to animals. In this world where rifles were displayed on gun racks visible in the rear windows of pickups, trucks hauling cattle densely packed into the available space plied the highways, and coyote carcasses hung on fences, people seemed very certain about the role of animals in human society. Save for household pets, domesticated animals were treated as mere modes of transportation and agricultural products. Wild animals were often considered pests, and killing them was a highly desired form of entertainment—although sometimes these same animals were held up as models of courage, strength, and stealth. There was little question about why animals existed—to serve human needs. If animals felt any pain, it was mostly irrelevant to how interactions were managed. While there was talk about the virtue of making a clean one-shot kill during a hunt, or the rapid slaughter of a steer, it seemed that these notions were about the skill of the hunter or butcher, not the ethical requirements for taking an animal life. Observing the different norms associated with the treatment and place of animals in this ranching-dominated culture, and then beginning to internalize them in my effort to fit in socially, was an important first step in the process of distancing myself from animals emotionally.

Nothing was more important in this regard than taking part in some activities that were, for the buddies who took me along, common and taken-for-granted aspects of rural West Texas life. Prominent among these friends was Charley Oates, whom I first met at a barbecue for new students. He stood out: tall and rangy, thin as wire, and wearing a large gray Stetson hat, western shirt with pearl snap buttons, blue jeans with a crease so sharp you could cut with it, alligator-skin cowboy boots, and a beautiful rust-colored vest (which I learned later he had made himself from a deer that he had killed on his ranch). I introduced myself so I could take in his costume up close. After shaking hands I asked him why he was dressed like a nineteenth-century cowboy, and besides, didn't he know that the West was already won?

I was joking, but the glare that I received declared that my joke had fallen flat. Slowly, however, a grin began to trace across his face, followed by a throaty laugh. He slapped me very hard on the shoulder and said that I would have to come to his ranch some weekend so that I could judge for myself whether the West had been won.

Charley was a "real" cowboy, not only in his dress but also in his life identification and purpose. He acknowledged that he was attending LCC at the behest of his grandmother, who was a generous donor to the school. She was hoping that he would develop intellectual and cultural interests; he, that he might meet a "good Christian woman" to marry and take back home to Balmorhea, Texas. Despite our obvious differences, we became friends immediately.

The first trip I took with Charley to his family's ranch in Balmorhea was both memorable and representative of the kinds of experiences I had with other new friends, including my roommate Bo and an agriculture major named Bob Pate. We left Lubbock on a Friday afternoon after our last class. After picking up a couple of six-packs of beer and some ice for Charley's beat-up Styrofoam cooler, we headed in a southwest direction. The little towns that we drove through looked like they could have been movie sets for the dozens of western movies that I had seen growing up. Among the cactus, buffalo grass, and irrigated farmland I saw oil pump jacks and drilling rigs. The sun was starting to set as we drove through the relatively big town of Odessa, Texas. We continued to Pecos, where we bought more beer, and then cut straight south on Highway 17, a narrow two-lane road, into Balmorhea.

It was dark now, and we drove directly to Charley's house, a stone and stucco structure. I was hurriedly introduced to his mother Halley and his father Jay. Halley smiled and said, "So you're the Yank? Let me hug your neck." She proceeded to fling her strong, slender arms around my neck and kiss me hard on the cheek. Then Jay moved forward, his hand extended, smiling broadly. They both appeared refined in contrast to Charley's rough-hewn cowboy appearance and manner.

After introductions were over, Charley emerged from the kitchen and handed me a large bean burrito. Burritos in hand, we headed out the door. On the way out, Charley picked up two rifles from a hall closet, handed one to me, and then threw me a couple of boxes of 22-caliber ammunition. The rifles were placed behind the seat, and we were off to find Johnny Kingston, on whose ranch we were going to work cattle the next day.

We found Kingston sitting in his late-model white Ford pickup truck, parked outside a café in the heart of the little town. I could see that he was drinking from a dark bottle as we approached. Charley cursed at him, whipping his hat like a Frisbee through the open truck window and knocking the beer bottle out of Johnny's hand and onto the seat next to him. Surprised and flustered, Johnny cursed back and opened the door of the truck so fast that it hit Charley straight on. Johnny jumped out, grabbed Charley around the waist, and tried to spin him backwards to the ground. Charley kept his balance and wrestled loose by prying Johnny's hands apart. Once separated, they circled one another, throwing punches that came close but never landed.

Laughing breathlessly, Charley offered to buy Johnny a beer to replace the one he had knocked from his hand, and the almost-real fight ceased. Johnny shook my hand and welcomed me to "God's country." His hand felt rough and viselike. Charley invited him to go rabbit hunting with us, but he declined, got into his truck, and drove off, purposely spewing gravel that rained down on Charley's truck.

I had felt myself swallow hard when the rabbit hunt was mentioned. After picking up another of Charley's friends, a friendly bull of a man named Kinley Moody, who jumped directly into the bed of the truck, we headed off to the hunt. Charley loaded and handed me one of the pump-action rifles as we drove through the desert brush. I was put in charge of a hand-held spotlight that was plugged into the cigarette lighter on the dashboard: I was to shine the light around the terrain, and when I spotted a rabbit to call out and keep the animal in the light while one of the others took a shot.

Kinley was an excellent marksman, typically killing the large jackrabbits with a single shot. As the night wore on and ammunition got low, rabbits that were wounded were either clubbed to death if they were conveniently close to the truck or left to die on their own, becoming easy coyote meals. Some of the rabbits that had been killed close to the truck were retrieved and thrown into the bed.

Eventually it was my turn. I had three choices: openly admit my desire not to kill, miss on purpose, or shoot to kill. Earlier in the evening I'd been able to avoid my turn by focusing on my spotlight duty—after all, I was getting quite good at it. By holding the light, however, I was already deep into the process of killing.

Charley must have perceived my reluctance: he began to talk about how much grass rabbits ate and how they deprived his cattle, and by implication his family. Kinley added that the rabbits lying dead in the pickup bed would soon be turned into food for his dogs.

These considerations helped quiet my wish not to kill. I no longer declined my turn. To be honest, this was not an intellectual decision. I just wanted to fit in, and if killing rabbits was what it required at that moment, I was willing to set aside my druthers. While I was finding the carnage a little disgusting, the beer haze and my desire to be one of the boys prevailed.

When I shot at and hit a rabbit, killing it cleanly, it jumped up in the air slightly like a sleepwalker being awakened. "Way to go, Yank!" said Charley, slapping my back and toasting my success with a beer.

As I bedded down that night on the couch in the den of Charley's house and sleep began to creep over me, in my mind I replayed the sleepwalker jump of the rabbits I had shot that evening. I had witnessed in real time how I could push away the sad and disgusted feeling by focusing on the praise I received for being a good shot.

The next morning Charley woke me up from a short but deep sleep. He had already made a pot of coffee and loaded his horse into a trailer that was in better shape than his truck. We headed off due south, into the Davis Mountains. As the sun rose and began to burn away a thin fog, I could see the incredible landscape all around us. As we drove away from the sameness of the lowlands of the Chihuahuan desert with its mixture of sand and sparse vegetation and gained altitude, we entered a universe of nature's diversity. Gray and tan boulders varying in size from a basketball to a elephant body were piled high and irregularly across the landscape. As we climbed higher, there were stands of ponderosa pine, piñon, aspen, and grasses mixed among undulating stream beds and vast, unhindered views. The rough terrain was touched with pink from the early sun, yielding a warmth that briefly glazed and softened the ruggedness. Charley laughed and asked, "Does this look tame to you, Yank?," gently mocking my stunned amazement.

When we arrived at the Kingston ranch house, a group of fifteen to twenty men was standing near the front porch. They all were dressed as if they had walked out of Randolph Scott's 1962 movie *Ride the High Country*. They talked quietly among themselves, some in Spanish, as they smoked, chewed, and spat and adjusted the saddles on their horses.

Charley and I went into the main house, where I was introduced to Duncan Kingston, Johnny's father. Duncan laughed and said that I was going to see a part of a steak dinner that few people ever got to experience. I told him I was ready. "We'll see," he said.

As I didn't know how to ride horseback at the time, I got to travel with Duncan in an open green Korean War–vintage jeep. The men on horseback headed off to round up the cattle, and several Mexican men went to the area where the cattle would be driven. There was no rush, so Duncan and I drove leisurely; he told the story of his family and the ranching business as he sipped from a pint bottle of Jack Daniels bourbon. He explained that the whiskey helped to relieve the pain in his knees. When we got to the large corral, there were already a few head of cattle in it. As the morning progressed, it became much more crowded as the cowboys continued to arrive herding small groups of animals. After a lunch of black coffee and tortillas, the branding process began.

A fire was built at the center of the open space outside of the corral, and a number of branding irons were shoved into its hottest part. If the calf being handled was small, it was roped and pulled close to the fire. A cowboy would then run alongside the animal, reach over its back, grab the loose skin on the opposite side near the underbelly, pick the animal up off its feet, then drop it hard to the ground in a maneuver called "flanking." Other cowboys descended, sitting on the calf while one jabbed a large hypodermic needle into its flank. For bull calves, another cowboy slit the scrotum with a pocketknife, stretched the testicles out with his free hand, then severed the cords with the knife — or sometimes his teeth. The horn buds were then dug out of the skull with a coring device. If the animal was more mature and had large horns, they were cut with a saw.

The calves bellowed loudly. Thin geysers of dark red blood shot out of the freshly cut horns of the large animals. and blood washed down the insides of their thighs.

Several of the calves had large, ugly wounds around their navels, full of undulating worms. These screw-worms were scraped out with a knife, and a glob of antibiotic cream was spread deep into the wound.

Finally, a cowboy carrying a red-hot branding iron stabbed the animal on its hip, holding the iron in place with his weight as the flesh sizzled and the burning hair and flesh produced an acrid smell. The final move involved cutting an ear-mark with a pocketknife into one of the ears. That completed, the loop of rope around the animal's neck was removed and it was released.

If the animal to be branded was large, one cowboy roped its neck while another threw a loop under its rear hooves and snagged its rear legs. The animal was then stretched between two horses until it lost balance and fell to the ground. At this point a cowboy dove between the animal's rear legs, putting his boots on one leg while grabbing the other with his hands. The animal's legs were then stretched apart as if the cowboy were trying to break an enormous wishbone. Once the animal was more or less under control, the rest of the procedures commenced. I tried the stretching move a number of times, receiving several hard kicks to the head and chest before I learned what to expect.

When the animals were released from their ordeal, they jumped to their feet, long elastic strings of mucus hanging from their noses and mouths, their eyes red with fear and pain. If the animal had been particularly hard to handle, it was often given a hard kick as a parting gesture.

Two piles were made away from the fire, one containing the testicles and the other the remnants from the ear-marks. I was told that the number of ear pieces equaled the number of animals worked that day, and subtracting half the number of testicles from the ear pieces equaled the number of females in the group. No pencils or paper were needed.

Throughout the afternoon the cows were pushed through the gauntlet with little rest for anybody. On occasion a bottle of Mexican mescal complete with a worm in the bottom, was passed around, and all the men took a hard drag on it when it got to them. The warm, wet burn cleared the choking dust out of my throat and relieved some of the tension. It was clear that this encounter was more than a set of procedures required by the agricultural necessities of growing meat. It was also men reminding themselves of a disappearing pioneer past, a test of toughness among tough men, and a fight between man and animal. No animal was permitted to escape without the marks relegating it to product status.

Later that night there was a big dinner of Mexican food and "son of a gun stew," whose main ingredient was the testicles cut out of the male calves. At first I avoided the stew; then I convinced myself that I might as well go all the way with this experience and consumed a hefty portion.

As we drove back to Lubbock on that Sunday afternoon, I felt exhausted and changed. I had made new friends, killed rabbits, branded cattle, and eaten strange meat. My left cheekbone and the bridge of my nose both had deep lacerations and bruising where I had been kicked by a struggling calf and punched by a drunk cowboy who wanted to play hard with the "Yankee greenhorn." I felt that I had done well and had helped to dispel the view of some of my hosts that New Yorkers were always weak, pale, and soft. In a sense I felt I had made the team.

My desire to fit in and prove myself was only part of the reason I crossed my own ethical barriers. Another part was the relief I felt in being freed from the fundamental ethical responsibility to relieve pain and oppression whenever it is possible and reasonable. The south Texas cowboys I had worked with lacked the burden of conscience; they didn't struggle with ethical questions, grieve over death, or become saddened by cruelty. They seemed to take the offensive against pain and suffering, confronting the cycle of life and death by participating in it directly and unselfconsciously, all the while maintaining a certain respect for the toughness and cunning of the animals involved. It felt good to try on this approach.

*   *   *

When I registered for classes at Texas Tech University in the fall of 1964, I still had no clarity about a specific major, only that it would be in a science area and related somehow to human health. Therefore I enrolled in a broad sampling of courses in biology and psychology. I was quickly struck by the content of my psychology classes. As a high school student I had become fascinated by writers such as Freud and Jung and the complex internal landscapes of the human mind that they described. Their view of the unconscious world that exists behind our daily everyday awareness had made good sense to me. And as an aspiring jazz musician, I felt that with music I was exploring that unconscious world and its emotional content. So I was totally unprepared for courses that portrayed psychology as a science of behavior unconcerned with the unconscious.

Not only did the course material not emphasize the dynamics of the hidden world of the unconscious, but the professors openly scoffed at and denigrated such notions. Talking about the internal unconscious states of human beings was characterized as akin to discussing the reality of Halloween ghosts. The "mental" life was seen as irrelevant to the development of a true science of behavior. People like Freud and Jung were portrayed as deluded scholars who wrote only quasi-scientistic fairy tales. In their place my professors offered up John Watson, Ivan Pavlov, and B. F. Skinner as the appropriate psychological pathfinders. Skinner's view, built on that of the early behaviorists, was that the only reasonable goals for the science of psychology were to predict and control behavior. Further, in pursuit of those goals, the study of behavior had to be limited to what was directly observable. So discussions about internal unobservable states like feelings and emotions were useless unless and until they became accessible to direct observation. Skinner argued that a neglected aspect of understanding behavior was the study of how the external consequences of behavior shaped the likelihood of future behavior. The question that needed to be addressed according to this paradigm was this: what rewarding consequences reliably follow the exercise of a given behavior, thereby maintaining it? The psychological treatment that logically followed from this analysis was the systematic removal of those consequences or the intentional rewarding of behaviors that would compete with the behavior in question.

From lectures and my reading I could see that psychologists like Watson and Skinner were about changing the world through the application of the principles of learning. Watson had famously boasted that appropriately applied behavioral principles could shape, not just influence, the outcomes of child development. He called for a direct and unsparing psychology of behavior unconcerned about the non-scientific sensitivities of the critics. Similarly, Skinner could imagine communities that were formed on the basis of his perspective. His novel *Walden Two*, which I read as a class assignment, was a serious attempt to illustrate the power of the laws of learning. These were thinkers whose ideas transcended the one-on-one orientation of the traditional clinical psychologist. They promoted the idea of a big psychology concerned not only with the individual patient but also with the structure and design of society itself. As Kathleen Kincaid put it later in her book about the *Walden Two*–inspired community of Twin Oaks, in Virginia: "We are still after the big dream—a better world, here and now, for as many people as we can manage to support. More, a new kind of human to live in that world: happy, productive, open-minded people who understand that in the long run, human good is a cooperative and not a competitive sort of thing. One man's gain must not, if we are to survive, be another man's loss."[1]

At first I didn't know what to think of this new paradigm. Were Skinner and the "radical behaviorists" asserting that this life that was going on under my skin, that was so real and important to me, really some set of epiphenomena that did nothing but entertain, and at times frighten? My first reaction was to resist these notions, thinking they must have been developed by people whose own feelings were for some reason stunted or inaccessible to them. But as the lectures proceeded and the passion and apparent reasonableness of the professors remained unabated, I found myself opening up to their arguments. The professors continuously offered up clinical case studies showing how altering the contingencies of reinforcement seemed to readily reduce the occurrence of peculiar and dangerous behaviors, leaving room for "normal" behaviors to replace them. In a similar vein, they showed that animals like rats, dogs, cats, and monkeys could be shaped to perform a wide variety of behaviors simply by adjusting the contingencies of reward. I read about how complex response patterns could be produced by manipulating reinforcement, without any need for internal thought processes, feelings, and intentions as explanations.

It was impressed upon us students that descriptions of behavior ought to be limited to that which could be directly observed. For example, one could not refer to a dog as "hungry," because this was an internal state; one could record only the "latency to consumption" when food was placed in front of the animal and refer to the number of hours since the dog's last meal.

During a course in physiological psychology in the spring of 1965, the professor, Patrick Strong, described how Yale neurophysiologist José M. R. Delgado had implanted in a bull's brain electrodes that could be stimulated with a radio transmitter, and then, in a public demonstration in a Spanish bull ring, had allowed the ferocious bull to charge him as he held only the transmitter. Within feet of reaching Delgado, Professor Strong told us, the bull had abruptly stopped, its rage "turned off" by the electrical stimulation in its brain.[2] Delgado was said to believe that neuroscience was only twenty-five years away from being able to apply knowledge of brain function to the treatment of psychological and neurological disorders as well as the modification of problematic parts of human nature, like aggression and greed. These discussions further suggested to me that psychology and neuroscience were on the verge of altering human society for the better.

One afternoon I was talking with Barbara Bass, a fellow student with whom I had struck up a friendship based in part on our mutual love of the poetry and music of the beat generation, but complicated by her obvious disapproval of my growing acceptance of the behaviorist perspective. Barbara mentioned that she had been babysitting for a new psychology professor named Elmer Davidson and his wife Jean. She suggested that I call him and see if he needed any help setting up his lab. "You two would be good for each other," she said.

I thought that this was a great idea, and I quickly followed up and made an appointment to see Dr. Davidson, who, I learned, had recently received his PhD from Penn State University and had studied under Charles Cofer, who was well respected and had written the standard text on human motivation. I saw this as a good recommendation for his competence.

As I sat down in his office, I described myself as a student just getting interested in research psychology who had no particular plan other than to get some laboratory experience. He described himself as interested in both human and animal learning and said he was in the process of building and equipping his laboratory. We left his office and headed toward his lab space, which was in the basement of the building. Except for some utility tables and a few metal folding chairs, the space was completely empty. We went back to his office, and he laid out his plans in more detail. He was going to develop the human learning side of his research first and would begin to construct the necessary equipment shortly; I would be welcome to participate. I asked about his plans for an animal research component, and he assured me that would be happening very soon. I stated my willingness to help.

Before I left, he told me I could use the small desk that in the corner to the right of his own desk, as I would be getting to the lab early and staying late once everything was operational. Then he reached into his desk drawer and flipped two keys to me. "You will need these," he said, "to get into the office and the lab when I am not here."

As soon as I left the psychology building, I called Barbara to tell her what happened. She was as surprised as I was about getting a desk and the key to the new professor's office. My career as a research psychologist had begun.

Over the next months, Dr. Davidson and I became very familiar. Fellow students teased me that we were joined at the hip. Together we built equipment out of scrap metal and plastic to test human memory, and we wired racks of electronic equipment to control the experiments. I learned how to use sheet metal tools and the basics of electronic relay programming, and was soon testing student volunteers in memory experiments. Sometimes at the end of the day we would go to a private club and drink beer. Barbara was correct: we were good for each other. I began to feel a level of confidence in myself that was very rewarding and a little unfamiliar.

As the semesters passed, thanks to my relationship with Davidson, I began to meet other members of the faculty, who began calling me by my first name. They asked how my research was going and began treating me as if I were already a graduate student. This new identity began to affect my study habits and my grades in a very positive way. I was reading faster and more intently than I had ever done before.

My work with Davidson was also an entry into two other laboratories. One was the lab of Dennis Cogan, another new faculty member, who had come from University of Missouri, where he had been a student of Melvin Marx, a well-respected researcher in the area of learning. The other was the laboratory of Sam L. Campbell, who had been a graduate student of B. F. Skinner while he was at Indiana University. Both were involved in animal research. Dr. Campbell studied the effects of punishment on learning in rats, and Dr. Cogan the effect of lesions in the hippocampus of rats' brains.

I began to spend my free time around Cogan's lab, as I wanted to learn more about both physiological psychology and learning. Dennis was a clever thinker and an encouraging mentor. As an introduction, I started to observe the brain surgeries done in his lab. I watched as rats were first anesthetized with the injectable drug pentobarbital. Once they were still, their heads were shaved and they were placed into a stereotaxic apparatus, where their skulls were stabilized by round metal ear bars and a plate that held the jaw in a fixed position. The pointed bars were driven into the ear canals until they popped the eardrums—a sound that was quite audible in a quiet room. Holes were drilled into the top of the skull. A suction device was then used to aspirate the tissue from the dorsal hippocampus.

The crudity of the procedures and the awkwardness of the student surgeons were quite obvious. I was surprised that after watching several of the surgeries I was offered the opportunity to assist and then, after doing that a couple of times, the chance to do a surgery myself. I was assured there was not much to it and this was the way to learn. "Watch one, do one, teach one" was the road to surgical expertise. I accepted the offer.

This was where I now believed I wanted to be, in an animal lab intervening in lives in serious ways and observing the behavioral consequences. While I liked the memory research that I did with Dr. Davidson, what we could do to the human subjects in the course of an experiment was too limited. With animals there were no limits, except for my own reluctance to injure in the name of science, and I was already wishing those limitations away. I was starting to feel the urgency to experiment. I needed to break through my residual reluctance if I was to make serious contributions to behavioral psychology.

As I recall, the male rat that was selected as my subject was used to being handled and offered no objection to being held by the loose skin on his back as I injected him with anesthetic. Soon his movements around the holding cage slowed down. Then he stumbled and lay down on his side, looking as if he were asleep. A pinch of the tail produced no flinch, so I was told to move on to the surgical procedures.

As I cut through the scalp, the rat shuddered. Obviously he was not yet fully anesthetized, despite the tail-pinch test results. We waited for a couple of additional minutes and then continued.

After finishing the procedure, I placed the still-unconscious rat back into his cage, laying it on top of a folded paper towel to protect it from the chill of the wire mesh floor. The uneven black stitches that held the scalp together oozed a bit of bright red blood. His chest barely expanded and contracted, but the rhythm was steady. I sat and watched the rat for the next half-hour. At one point he raised his head slightly and blinked. All was going well with the recovery phase, so I accepted an invitation to get a cup of coffee in the department lounge. When I returned to check a short time later, the rat's body was still and his chest no longer moved. He must have died shortly after I left for the break. In that moment I realized that a life had been lost because of my ignorance and clumsy technique. Horrified and embarrassed, I thought I should never have been allowed to try this procedure.

I reported the death to the lab supervisors. Their collective judgment was that I had aspirated too much brain tissue when I was trying to remove the dorsal hippocampus. They assured me this outcome would change as I got more experience and learned to better recognize the anatomical landmarks. It was suggested that I do another rat as soon as possible.

I asked how to get someone to examine the rat for the actual cause of death. I was told that the rat was "an extra" and to simply put it into a plastic bag and throw it into the garbage out behind the building.

"That's it?" I asked again.

"Yes, that's it," one of the supervisors answered, looking surprised at my inquiry. "It's just a rat, for God's sake."

I decided that I would keep doing the surgeries until I could get the rats at least to survive. It was a skill that I wanted to add to my list of capabilities. As time went on, I found that my scalpel cuts, lesioning technique, and suturing did improve. Nonetheless, I carried many more bags of "extra" rats to the trash bin before one finally survived long enough to be tested. The advice I received was not to quit and to keep trying, and eventually I would discover what was necessary to be successful.

Around this time I had begun seeing a bright, quiet young woman named Karen who was a home-economics major from Houston. On our second date I proudly described my research experiences in the department of psychology, thinking that she would be particularly impressed by my surgical exploits. Instead, she responded with revulsion.

I was not exactly stunned by her reaction, as underneath it all I was a bit horrified myself. She could not believe that my fellow psychology students and I could just declare ourselves to be animal neurosurgeons and commence to practice on any available beast. In what would have amounted to a rage for this gentle person, she claimed that she got better supervision when she tried out bread recipes in her cooking class. Instead of acknowledging her point, I put on the blasé attitude that I saw in my lab partners.

Over coffee late that night, I relayed her criticism and my lab friends scoffed at her emotionalism. One of the grad students later sent me a note with a quote by behaviorism's founder, John Watson: "The raw fact that you as a psychologist, if you are to remain scientific, must describe the behavior of man in no other terms than those you would use in describing the behavior of the ox you slaughter, drove and still drives many timid souls away from behaviorism."[3] The word *timid* was underlined. The message was that scientific progress required a degree of ruthlessness.

Yet Karen's outrage continued to bother me. It brought to memory my own reaction when my cousin Robert and I encountered the boys on Long Island sticking a knife blade into the live eel to see what it looked like. At that time in my childhood, I saw with moral clarity that we had no right to access such knowledge in that way. Things were different now. There was so much at stake that required action, and the side effects of harm in animals just had to be endured and then hopefully ignored. This was the transformation that was required. It was also necessary, it seemed, to shroud what was taking place in the lab from the "timid" and uninformed in order to protect the critical process of research.

I began to expand my hands-on animal research to the department of biology. While taking a course in comparative anatomy, I asked the professor if he had any ideas for an extra-credit research project that I might do. He said that there had not been much work on the neuroanatomy of the blacktail prairie dog, an animal that was quite abundant in the area around Lubbock. He suggested that I might trap a few, euthanize them, and then, under his supervision, dissect the brain, cranial nerves, and spinal cord. If things went well, we would write a paper describing the results.

After considerable effort, I managed to capture a single male prairie dog on the outskirts of town. Instead of following through with the original plan of euthanizing the animal with chloroform in the lab and then freezing the carcass until I got together with Professor Rylander, I took the animal back to my apartment so that I could take some time and watch his behavior. As I had been cautioned about personalizing experimental animals, I objected when my roommate Dale Robbins named him Bernard. While the prairie dog lived with us over the next few weeks, I steadfastly refused to use the name. I thought of myself as practicing necessary scientific objectivity. The animal adjusted rapidly to urban apartment life, and Dale grew very attached. He fed Bernard fresh carrots and apple slices and talked to him in that high voice that is reserved for babies and family pets.

Right before the end of the semester, Professor Rylander asked me if I had trapped a prairie dog that we could dissect. I reluctantly admitted that I had indeed trapped one, but that rather than lying frozen in the lab he was now alive and staying in my apartment. At Rylander's insistence, we made a date the following week to euthanize the animal so that we could begin. The day before that appointed time, I was finally able to find the courage to remind Dale of Bernard's purpose.

"You're still going to do it?" Dale asked incredulously. I defended myself, pointing to the contribution to science the dissection would conceivably make and suggesting that it trumped Dale's attachment. I knew at the time that this comparison was fallacious given my low level of expertise in neuroanatomy, but I was trying out the justifications I had heard and did believe it was true in the general sense.

This argument was different from the one I had had with Karen. Her concern was that my laboratory partners and I didn't know what we were doing and as a consequence were wasting lives and time. Had she been satisfied with my level of supervision and knowledge, I think she would not have objected—at least not so strenuously—to my involvement with animal research. Dale was arguing about the value of his emotional attachment and how that had to be factored in to the decision about Bernard's fate. In other words, he claimed that his feelings and concerns were relevant to my decision.

What didn't come up in our disagreement was the animal's point of view. For both of us, his removal from his environment and the subsequent disruption of his life was irrelevant to the debate. The animal existed in our lives for us: for Dale, Bernard was as an object of affection, while for me he was an object of investigation, a means to the end of gaining information about the structure of the central nervous system of a burrowing mammal.

I was determined to have it my way even though I felt affection for Bernard and reluctance to sacrifice his life for a project of questionable value. The matter had come to be a test of whether I could face the grim task of killing this animal so that the project could begin. The question was, could I put aside my own emotional objections and proceed in spite of them? I was coming to appreciate that this ability, call it "selective ruthlessness," was a necessary skill in the profession I was pursuing.

The next morning I went to the Psychology Department animal facility to make certain that there was an ample amount of chloroform available to euthanize the animal and formalin to preserve his body. When I returned to the apartment around midmorning in order to retrieve Bernard and get on with the task, I could find neither Bernard nor Dale. Dale later denied releasing Bernard on purpose, insisting that he had taken the prairie dog out to graze on the lawn for a few minutes and that when he went to get him, he was nowhere to be found.

Unfortunately, Dale's attempt to convince me that the pursuit of science had to be balanced against other values like emotional concern for another being was mostly lost on me at the time. When I went to the lab section of the anatomy course the following week, I found fresh prairie dog carcasses set out at the lab stations along with a prepared written exercise focusing on dissecting the trigeminal nerve. Apparently these animals were easily available through lab suppliers. I wondered whether Professor Rylander had sought to steel my sensitivities by instructing me to carry out the trapping and euthanasia by myself. Today I have no doubt that was the case.

∗   ∗   ∗

By the fall of my senior year, I had decided for certain that I would apply for graduate school. I had become convinced that a research career in behavioral science was a worthy goal. In psychology the opportunities for study and research seemed limitless. I loved the feel of working in the lab, conducting experimental procedures on human and animal subjects. As trivial as it sounds, I liked having keys to locked rooms. It symbolized the importance of my research activities. I liked the fact that I was trusted to handle dangerous drugs, conduct surgery, build equipment, and statistically analyze the data that was collected. I was intellectually moved by the arguments about behavioral theory in which I participated and found that my passion would rise during the debates with an intensity that surprised me. I liked the late-night meetings at the Country Inn, where a number of us ate scrambled eggs with hot sauce and drank gallons of coffee and iced tea as we argued about the meaning of our research results and imagined the real human problems we might eventually solve. We talked about finding solutions to learning disorders, depression, brain injury—all looked like they were within reach.

I had become confident about the knowledge the experimental method could reveal so long as our designs were courageous and not "timid." This reinforced the original notion that in pursuing neuroscience I could end up contributing something relevant to the understanding of Parkinson's disease, from which my father was suffering terribly, or perhaps something about anxiety and depression, which plagued my sister and grandmother.

The experience of developing a question about some aspect of theory, designing an experiment, and then having an animal's response to the experiment declare whether I was correct, or at least on the right page, was beyond exhilarating. The price that the animals paid for helping to clarify the value of my questions began to pale in the face of the scientific progress I believed I could see. And perhaps more importantly, the emotional understanding that a price was being paid in the first place began to fade, replaced by an abstract, intellectual one. It was as if the rats, cats, dogs, and monkeys were there just for us. They had no lives other than for making research possible. Being in the position of making decisions about how their lives would be employed left me feeling privileged, special, and on the verge of understanding things of importance.

I spoke with the faculty members with whom I had established relationships. They agreed that the extent of my research experience would be seen as a very positive asset at most programs and would raise significantly my chances of admission. Dr. Davison suggested that I seriously consider the University of Oklahoma, where it was possible to get both PhD and MD degrees. Other faculty felt that I would be better off focusing on experimental psychology, where they believed I had demonstrated real promise. They assured me that they would write strong letters of recommendation regardless of the option I selected. As the year wore on, I gravitated toward the psychology graduate school option but could not sort out the question of which programs to apply to. In the end, I decided on the simplest route: to stay on as a graduate student at Texas Tech.

The faculty members were good, I got along well with them, and I knew my way around. These were all considerable advantages.

By the spring of my first year as a graduate student, however, I felt I was in an educational holding pattern. I began to appreciate that the facilities available for the conduct of animal research were very limited. The facility had space for only a few singly caged monkeys, a small section for cats, and some pigeon cages that were at that time occupied by five red-tailed hawks being trained to kill pigeons around the campus. The remaining available animal housing space was devoted entirely to rodents. This limited variety of research animals did not fit with the changing ethos in psychology, which pressured researchers to study more than just rats (and college sophomores) if its inferences about normal or abnormal behavior were to achieve useful credibility.[4]

Educational issues aside, the summer was coming and I had no realistic prospects for financial support, and that fact was very much on my mind. The thought of perhaps returning to New York, to look for a summer job and live with my parents, seemed like a backwards step I wanted very much to avoid.

One afternoon in late April, I was lamenting my circumstances to my friend Kenichi Takemura. Ken was an advanced graduate student in the Department of Psychology who was as good a listener and problem solver as any clinician I ever knew. That afternoon we were relaxing in his small apartment drinking his home-brewed beer. Ken well understood the conflict in which I was embroiled, both conceptually and practically. Before coming to Texas Tech he had been a graduate student at the University of Wisconsin under the mentorship of Harry F. Harlow. Ken was interested in comparative psychology and in nonhuman primate learning in particular, and Wisconsin was the place to pursue those interests. After completing his undergraduate degree in psychology in his native Japan, he had been admitted to the graduate school of the University of Wisconsin to work with Harlow at the Department of Psychology Primate Laboratory.

For several years he had progressed well as a student and researcher and was in good academic standing with a promising academic future. All this changed when he took his preliminary exams. The grueling pace of the testing schedule and the failure rate for these exams were notorious at Wisconsin. The tests strained the abilities and anxiety tolerance of all who took them. The examination committee deemed irrelevant the fact that English was not Ken's primary language, and they refused to allow him any extra time.

Ken told me that no matter how hard he tried, he could not compose a response and write fast enough to answer enough of the questions to gain a passing score on his major exam. So because he could not be advanced to doctoral candidacy, he had to leave. This was an academic tragedy of a sort not uncommon at Wisconsin. He applied to several other universities, and Texas Tech accepted him on the strength of his publication record and a glowing recommendation sent by Harlow.

As the afternoon passed, the warm, sweet beer lowered our inhibitions and increased our courage. By the time the sun had set, Ken offered to call Harlow directly and ask if there was a spot for me as a summer intern at the Wisconsin Primate Lab. I knew this call was going to be difficult, as he confided that he believed he had embarrassed Harlow and disgraced himself by his failure. His willingness to call was a sign of support and real friendship, and I appreciated it immensely.

Later that night, after I had recovered a more realistic frame of mind, I had extreme doubts that such a possibility would actually materialize. After all, I was at Texas Tech, a good B-level institution at the time, and Wisconsin was among the rarefied elite in psychology.

A few weeks after having had the conversation with Ken, I was absolutely surprised to receive a personal letter addressed to me with a University of Wisconsin Primate Laboratory return address. The brief letter offered me a half-time position as a research assistant for ten weeks beginning the first of June. The last sentence was hand-written in blue ink and read: "Although this position will be funded at a half-time pay rate, the expectation is that you will work full time—at least." It was signed "Harry Harlow."

While this condition may have been a formal violation of some employment law, I saw it as simply indicating the seriousness with which research was pursued at Harlow's laboratory. I felt challenged by the directness of the expectation. The law had been laid down: research progress had the highest priority, and my nonresearch life was irrelevant.

*   *   *

I arrived in Madison in late May, almost a full week before my formal term as an intern was to begin, and within a couple of days I had found an apartment only a few blocks away from the Primate Laboratory. On the Monday I was to report, I walked to the lab and arrived there a little before seven. Standing before the doors, I endured a brief but powerful crisis of confidence and then entered the unlocked building. I was immediately greeted with the familiar smell of concentrated animal life, a warm, humid odor deeply infused with body oil, soap, and waste.

After I climbed a short staircase, a large reception area with an adjoining office appeared immediately to my right. I looked into the room but saw no one. To my left were two closed office doors, one with a black-and-white sketch taped to it. It was a picture of two chimpanzees copulating. The image was quite strange: The male chimp was poking his eye with his right index finger and apparently waving or saluting with his left hand. Instead of exhibiting the species-typical mounting posture, he was seated awkwardly. The picture made no sense to me. I would soon come to understand that the picture was a kind of symbolic shorthand for the bizarre stereotypical behaviors that resulted when chimpanzees were raised in socially isolated environments.

I continued down the hall and came to what looked like a lunchroom with four nondescript gray Formica-topped tables, each surrounded by four to six chairs. At the table closest to the far wall sat a man in a blue Hawaiian print shirt, holding a white typewritten page in front of him. The stub of a cigarette smoldered in a round glass ashtray to the right of a pile of papers. A stained paper cup half full of coffee the color of volcanic glass sat precariously close to the manuscript pages. His face showed more than one day's growth of

beard, and the lenses of his glasses looked smudged. He nevertheless maintained something of an aristocratic air. I assumed that he
was a faculty member or some type of senior administrator. As I
approached, he looked up slowly as if he had a stiff neck. After a
moment, a small but discernible thin-lipped smile broke the seriousness of his focus.

I introduced myself, told him that I was a newly arrived summer
research intern, and asked him if he could direct me to the office of
either Ken Schiltz or Helen Lauersdorf. He took a deep drag on what
was left of the cigarette, producing a hissing sound as he inhaled air
around the loose cigarette. He exhaled, fixed his gaze on my face,
extended his hand, and said, "I'm Harry Harlow. I didn't catch your
name."

His voice seemed to arise from deep in his throat. His hand was
cold, his skin soft, and his grip firmer than I would have expected. He
asked me to sit down. He appeared relaxed but said nothing further.

I couldn't quite take all this in. Five minutes earlier I'd been questioning my suitability to be there, and now I was face to face with one
of the most influential psychologists in the history of the modern
discipline.

After waiting awhile for him to speak, I broke the silence, explaining that I was from Texas Tech and a colleague of Ken Takemura's. I
continued that I understood that Ken had called him about me and
that he had been gracious enough to arrange a summer internship.

His eyebrows lifted noticeably when I mentioned Ken's name. After smoking the better part of another cigarette in silence, he said that
I had already passed Helen's office and was now only a short distance
away from Ken's. "One door further down the hall, past the time
clock on the right. His name is on the door. I doubt that even a Texan
could miss it," he said. I couldn't tell if he was smiling. "Remember,
Ken Schiltz can teach you a lot about how to test monkey learning,"
he continued. "He is the second-best person in the world at it."

As I slowly got up to leave,, he spoke again. This time his voice
stiffened into a louder declarative tone, and he rose up a little in
his seat. "Ken Takemura is an excellent man and don't you forget
it." He then relaxed, and his gaze returned to the pages in front of

him. I could tell that our conversation was over. I simply nodded in agreement, excused myself, and left.

This first encounter with Harry Harlow was in a sense a microcosm of the many interactions that I would have with him until shortly before his death in 1981. Basically friendly, totally self-contained, and sociable, he always displayed a sharp edge tinged with sarcastic humor. Discussions with him were always full of silences and cigarettes. Despite the gruff exterior, he was deeply loyal to and protective of his students and staff.

When I got to Ken Schiltz's office, I found the door open and a slender balding man sitting behind a rather cluttered desk. His smile was broad and honest. After I introduced myself, he stood up quickly, leaned forward, and energetically shook my hand as if he was honestly pleased to meet me. We talked for over an hour, touching on monkeys, Harlow, the variety of research projects going on in the lab, Wisconsin Badger football, and the Green Bay Packers. He introduced me to his coworkers as they arrived for work. Before I left his office I had been given advice about where to shop and eat, recruited to play on the lab softball team, and directed to Johnny Laugen's Bar, the unofficial lab hangout. Ken exuded an air of helpful friendliness that was obviously authentic. I would soon find that this warmth, good humor, and openness were characteristic of most of the staff, students, and faculty.

During the next few days I received a thorough orientation to both the Department of Psychology Primate Laboratory and the Wisconsin Regional Primate Research Center. The Psychology Department Lab was a converted cheese factory with an extensive new addition. The Regional Center, just across a small alley, was all new and of more modern construction. The lab comprised animal holding and testing rooms, food preparation and medical treatment rooms, a full surgery, a large shop area with full-time carpenters and metal workers, an electronics shop, a photography unit, administrative offices, and a set of chemistry labs. I was awe-struck.

My orientation covered all aspects of basic primate care, including proper housing, transporting, handling, and feeding. I was shown how to clean the apparatus I would use in the testing and was made aware of the monkey diseases I would need to avoid. I found that monkeys at the primate lab were usually housed individually in stainless-steel cages with mesh fronts, ceilings, and floors and solid side walls. The cages were arranged in rows typically two tiers high. Under the banks of cages ran a sloping tray that caught feces, urine, and the remains of uneaten food. Several times a day a technician would flush this debris down the drain with a high-pressure hose. Several times a week, the monkeys would be removed from their cages and held in transport boxes while their cages were cleaned with pressurized hot water, soap, and disinfectant.

At Texas Tech monkeys were given a daily measured portion of a commercially available chow, and that was that. At the Wisconsin Lab feeding was a much more elaborate operation. First, each monkey was given a "sandwich" made up of a brown sticky spread containing vitamin supplements and a dose of the antituberculosis drug isoniazid lathered generously between two slices of wheat bread. Next, several pieces of fresh fruit were delivered. At Tech we rarely provided fruit to monkeys, and when we did it was castoffs from a grocery store. Here the fruit was the best that was available for human consumption, and it was provided each day. Finally, a measured portion of monkey biscuits completed the meal. Care was taken to ensure that adequate water was available in the large bottles attached to the cages.

Ken Schiltz introduced me to the Wisconsin General Test Apparatus (WGTA), a device that Harlow and his students had developed in the 1930s and that was still very much in use,[5] and provided a short course on testing procedure. The apparatus can be conceptually divided into three parts: a subject-holding section, a problem-presentation and animal-response space, and an experimenter preparation and observation section. The subject-holding section was simply a cage constructed to limit the overall motor activity of the monkey subjects. The front was composed of vertical bars instead of mesh so that the animal could easily reach his or her hands through. A movable opaque screen separated the animal-holding cage from the problem-presentation space. The screen prevented the monkey from observing the setup of a problem and responding to it before all was ready. The problem-presentation area consisted of a tray or form board upon which problems were placed. The tray moved on a track that could be moved by the experimenter in or out of the animal's reach in order to control response opportunity. The experimenter section was partitioned from the presentation and response space by a one-way mirror that permitted the experimenter to observe the animal's response.

The types of tasks presented in the WGTA were intended to test discrimination ability, interproblem transfer, memory, and various kinds of concept learning. The stimuli used in the WGTA for these tests were constructed of "junk" three-dimensional objects such as small toys, blocks, and irregular pieces of plastic that were attached to bases sufficiently large to cover the food wells that were cut into the form board tray. In simple discrimination tasks, the objects differed from one another on many dimensions like size, color, and shape. To set up two-choice object discriminations, the experimenter consistently placed a food reward beneath the randomly selected "correct" object, regardless of its position on the form board. The experimenter would run a trial by raising the opaque screen and sliding the tray toward the monkey, permitting it to choose one of the objects. The choice response required only that the monkey move one of the objects in order to reveal whether there was a food reward beneath it. If the monkey chose correctly, he or she would pick up the reward and eat it immediately or place it in a cheek pouch to be consumed at a later time. Then the opaque screen would be lowered, and the experimenter would retract the tray and set up the objects for another trial.

Harlow had shown early in his career that if monkeys were give many hundreds of such problems, each presented for only six trials, they began to develop general strategies, or what he called "learning sets," that were applicable to solving all the problems regardless of the specifics of the stimulus objects. In Harlow's terminology, the monkeys "learned how to learn."[6] More specifically, he believed that the monkeys tested various solution hypotheses rather than focusing only on the characteristics of a given problem. This notion was quite revolutionary, because it drew a picture of the mind of the monkey quite different from the favored behavioristic assumptions of the time.

After I received instruction on how to maneuver the WGTA and set up the various commonly used problems, Ken gave me the ID number and location of a monkey not currently on any experimental protocol with whom I could practice. The monkey lived in a windowless basement room along with about twenty other adult rhesus monkeys. As I walked down the rows looking for my monkey's number on the food box, some monkeys screeched, stood up, and swayed back and forth on their back legs while gripping the wire mesh on the front of their cages. Others glared at me and made loud hooting sounds as they bobbed their heads. Some rattled the hinged front walls of their cages. A few retreated to the very back of their cages, wrapping their arms around themselves as if in abject fear.

I found my monkey in the top tier of cages in the middle of the room. He was a large grayish-brown male adult who, according to his records, weighed sixty-plus pounds. His powerful appearance gave me pause. For about fifteen minutes I tried unsuccessfully to get the monkey to enter the transport cage. I began to reflect on the virtue of doing research on rats—you just had to grab the animal by the base of its tail and lift it into whatever apparatus you wanted. Finally, after I left the room briefly to allow a fresh start, the monkey complied and entered with transport cage without any fuss.

On the second floor of the lab there was a series of WGTA testing rooms along a hallway illuminated only with a few red bulbs. I found an unscheduled room and entered with my monkey. I raised the transport cage to the open door of the subject-holding cage of the WGTA. The monkey balked for a few moments and then walked into the cage. I closed and latched the door and went to the experimenter chair behind the one-way vision screen.

Now I could set up the first problem, which used a small red metal truck and a yellow plastic ball that had been cut in half. I randomly selected the truck to be the correct stimulus and placed a raisin in the food well beneath it. The yellow half-ball covered an empty food well. I raised the opaque screen so that the monkey could see the stimuli. He immediately moved to the front of the cage and took a look at the objects. I then slid the tray with the objects within his reach. He reached toward the plastic ball and flicked it off the covered food well with the back side of the fingers on his right hand. Seeing the empty food well, he stood up immediately and began to circle to his right. I retracted the tray so that he could not respond again and lowered the opaque screen.

Now I rearranged the position of the correct truck object according to a predetermined sequence that was printed on the data recording sheet. Raising the screen again, I pushed the tray in toward the monkey. In an instant he pushed the truck aside, retrieved the raisin, put it in his mouth, and began another right-hand circle inside the cage. And so it went for a dozen separate problems. If he guessed correctly on the first trial, where the possibilities were 50–50, he stuck with that object regardless of its position on the five succeeding trials. If he guessed incorrectly on the first trial, he switched to the other object on the following trials. At the end of the dozen problems, he had made only three errors.

I then set up and ran a series of delayed-response tasks that involved showing the monkey under which of two identical blocks I was hiding the food reward, lowering the opaque screen for a set amount of time to hide the objects, and then raising the screen and sliding the tray forward to allow him to make his choice. He was very good at this type of problem as well, making correct choices more than 90 percent of the time.

For the next two days I practiced testing the monkey on a wide variety of problems and began to feel that he and I were developing a kind of rapport. When I arrived each afternoon at the same time for our session, the monkey would be waiting for me at the front of his cage. He would give a low grunt that seemed like a greeting, smack his lips at me, and climb into the transport cage without hesitation. I found myself calling him Al, as he reminded me in some way of my uncle Al, a gruff but warm person whom I always enjoyed seeing as a child. I was audacious enough to add the name Al in quotes to the white ID tape on his food box. I don't know why that move seemed important, but it did.

I was fascinated by the monkey's apparent involvement and

interest in the testing itself. After making an error he would often grab at the wire mesh floor of the holding cage with his hands, as if expressing frustration. If I inadvertently failed to lower the opaque screen all the way, I could see him flattening himself on the cage floor and peeking through the crack to observe where I was placing the rewards. Since he was not food deprived, something other than the need for food was motivating this monkey.

These impressions were not figments of my imagination, for when I brought them up with experienced monkey testers, they shared similar stories. They readily spoke of having been "fooled," "deceived," or otherwise "tricked" by monkeys. Their stories were full of obvious respect for the monkeys' spirit and sometimes awe for their apparent intelligence. Like them, I found myself describing monkeys as "perfectionists" or saying they "wanted to beat the game and fool me." I surprised myself with this kind of thinking, as it violated the strict behaviorist training I had been receiving in graduate school up to that time. That training told me to limit my descriptions of behavior to what could be objectively seen and not to impute or infer the existence of mental states not possible to observe. This training also warned that to anthropomorphize the behavior of animals with human descriptive concepts, even nonhuman primates, was a scientific sin punishable by excommunication from the church of Skinnerian psychology.

For the rest of the summer I tried to clean up my language and to be consistent with these "hard-nosed" values. I studiously replaced *hungry* with *food deprived, fearful* with *avoidant,* and *smart* with *accurate.* I even found myself criticizing the sloppy language of some of my workmates. I felt I was taking an intellectual stand. One afternoon after I had been at the lab for about a month, I went back to Al's cage and blacked out the name on the ID tape.

With my orientation complete, I was ready to involve myself for real. I was assigned to work with a visiting faculty member, Sheo Singh, a professor of psychology at Panjab University in Chandigrah, India. He had developed prominence as a primatologist studying the behavioral differences between urban and rural monkeys in India with respect to their learning ability and social behavior.[7] When

Professor Singh and I met for the first time, he wasted little time in describing the experiment on which I would be working. His plan was to study the effect of the drug scopolamine on two-choice discriminations tested in the WGTA. He hypothesized that at higher doses the drug would interfere with short-term or working memory, thereby increasing the number of errors even in experienced monkeys. I asked how he became interested in this problem, and he explained that in India he had done similar studies with humans and found interesting results. He complained that the degree of control that he could exert on the human subjects regarding their diets and work cycles left any firm conclusions out of reach. I wondered but didn't ask how he'd managed to get humans to volunteer for such studies.

At the end of the meeting he gave me the list of monkeys that would be used and the schedule of the drug doses. I was surprised to see that one of the monkeys on the list was the one with whom I had practiced for the last weeks. This illustrated how the monkeys in the lab were used over and over again by different researchers throughout their lives. Surprisingly, there was no discussion that I heard about how those serial experiences affected the results of any one study, and certainly no talk about whether there was a cumulative negative emotional impact on the monkeys. I deflected these concerns rapidly by assuring myself that the monkeys most likely enjoyed the stimulation of being tested as opposed to sitting day after day alone in their single cages.

Testing the monkeys for Singh's research was similar to what I had done with Al during my orientation except for the added step of injecting each monkey with the drug or the sham prior to testing. This was not particularly easy, as it involved restraining the animal in an apparatus called the iron maiden. Since this apparatus bore a strong resemblance to the medieval torture device of the same name, it was not surprising that the monkeys resisted mightily. Al, for example, immediately recognized the purpose of the apparatus and would not move from the transport cage into the squeeze portion of the apparatus until forced to do so by the coordinated efforts of the animal handler Laverne Rossler and me. At first Singh did the injections, but his technique was so rough and unsteady that eventually I suggested that he permit me to do the entire procedure with Ross's help. He seemed relieved by my suggestion and agreed.

Since Dr. Singh was first looking for an effective dose, he started low and worked higher in sequential fashion, interspersing sham days from time to time to reduce the possibility of inducing cumulative drug effects. The data showed no evidence of poorer performance on any of the drug days. Then things changed dramatically. After administering a new dose to Al, I transferred him to the WGTA as usual and then timed the standard thirty minutes before starting the testing. When I returned to the test room, Al was lying on his side on the cage floor in a rather awkward position, breathing rapidly but evenly. With some difficulty he eventually sat up. His back legs were splayed open, and his hands and arms formed a rubbery kickstand in front of him. He took a long drink of water from the bottle attached to the cage. The water then dripped from his drooping jaw and lower lip.

I paged the veterinarian, and he returned the call promptly. He indicated that what I was seeing was the effect of a high dose of the scopolamine and not to worry, as it should wear off in an hour or so. When I returned to Al, he was sitting with little more stability. I decided to see if he would work to solve any problems.

I set up the problems and began. When I raised the opaque screen for the first time, Al pulled himself to the front of the cage, looking like a drunk bellying up to a bar. He looked at the objects even as his head wobbled. He reached unsteadily through the bars and made a choice. His response was correct, uncovering a food well containing two pieces of dried corn. He made an attempt to pick up the corn, but the kernels fell from his hand and rattled into the metal drop pan beneath the cage. He stared at the kernels beneath him as if he was amazed to see them there. I ran several more trials, and each time he showed an inclination to respond, but his lack of coordination made it impossible to complete the task. I paged Singh and described what was happening. He instructed me to continue.

It took several hours, but I ran all the monkeys for as many trials as I could. I definitely felt uneasy. It was clear that the monkeys were

having trouble seeing, maintaining balance, and eating the food rewards they earned, but they did continue to sporadically perform.
As I think about the episode now, it seems quite pathetic. If these
had been human subjects in front of me, I would surely have asked
if they were feeling nauseous, high, dizzy, embarrassed, or perhaps
humiliated. In this case the vet had said there was no reason to worry,
and Dr. Singh wanted me to continue. So I did.

We were able to complete the study before I had to return to Texas.
I was skeptical that much could legitimately be concluded about the
effect of scopolamine on memory in monkeys based on the way we
tested it. How was it possible to distinguish scopolamine's effects on
motor coordination and vision from its effects on memory per se? In
any event, at the time I found the experience of working and talking
with Professor Singh rewarding and learned a good deal about testing, restraint, and drug dosing techniques. I never saw any evidence
that he published the results of the study.

During the summer I also worked in the laboratory of V. J.
Polidora, who studied discrimination learning, drug effects, and the
automation of learning test apparatuses. I was assigned to work on
a project involving the study of aversive conditioning using a device
called the shuttle box, a long, narrow rectangular chamber with a
floor constructed of stainless-steel bars that ran perpendicular to the
length of the chamber. The basic task required the monkey subjects
to learn that when a particular stimulus—either a light of a specific
brightness or a noise of a certain decibel level—was presented, they
had only a few seconds to run (i.e., shuttle) to the opposite side of
the chamber to avoid the application of a painful electric shock or a
blast of high-pressure air. If the monkey failed to shuttle during the
safe period, it was shocked or blasted until it escaped to the other
side of the chamber. This basic protocol was similar to that of thousands of other experiments studying aversive conditioning using
other species, mostly rodents, as subjects. The proposed unique
contribution of the project to which I was assigned had to do with
the use of primates as subjects and comparing the effectiveness of
air blast and electric shock.

Testing with the shuttle box was very different from doing tests with the WGTA. With the latter, you interacted directly with the monkeys and could get a sense of the animals as individuals. In these experiments, the testing was completely automated. Once a monkey was in the chamber, the doors were secured and the programming equipment was turned on. When the program automatically ended the session and recorded the data, the tester was signaled to remove the monkey and return it to its home cage. While in the chamber the monkeys could be viewed via a closed-circuit television hookup. The black-and-white picture was grainy and lacked detail, but it made it possible to track the gross-motor movement of the monkey. The setup seemed very precise and sophisticated to me, and I liked working with it.

I was put in charge of testing a squad of adult stump-tailed monkeys who were in the air-blast condition. From the beginning, it was clear that receiving a strong air blast to the body or face was a powerful motivator. When a monkey was first exposed to the air blast, it immediately startled, jumped vertically—often hitting the top of the chamber—and defecated heavily. Understandably, the monkeys quickly learned to run to the safe side of the shuttle box with a high degree of accuracy.

The completely automated nature of the test situations created distance in the relationship between the monkeys and the testers. At the time I thought this was an important scientific advance given the loose level of anthropomorphism I had been seeing. It also had its disadvantages.

One day I was running a stump-tailed subject who had been acquiring the avoidance response pretty well during the previous sessions of testing. I became aware that he was starting to neither avoid nor escape on a consistent basis but spent an increasing amount of time at one end of the chamber. The TV camera revealed the animal to be hunched over, as if he were sleeping. While I thought the situation was peculiar, I knew learning was not just a simple linear process, and the air blast would likely affect some monkeys more negatively than others. Perhaps for this monkey the effect was so highly fear-provoking that it interfered with the learning process.

I continued the session. This freezing response went on for the better part of an hour. Then, as I glanced at the image on the TV screen, I could see that the monkey was not only looking directly into the lens of the camera but was also grimacing, with lips pulled back and teeth exposed, making a chattering movement. It was an obvious fear grimace. Something was wrong.

I went into the sound chamber to check the equipment. I found that a position sensor was malfunctioning, causing the monkey to be punished with air blasts regardless of whether he shuttled to the correct side of the chamber or not.

I placed the transport cage up to the guillotine door to remove the monkey so that I could begin to repair the problem. When I opened the door, the monkey shot into the cage with such force that I nearly lost my balance.

I told Wayne Thompson, the lab foreman, and he joked that the Badger football team could use a fullback like this monkey for the upcoming season. The joke relieved the guilt I felt in allowing a malfunction to bring significant distress to the monkey. Later that night, I wondered if the monkey had been intentionally calling for help as he stared into the camera and grimaced.

In addition to learning a great deal about primate research that summer, I was fully included in the social life of the lab. At noon and at other break times the lunchroom was full of people, conversation, political argument, and excited games of bridge and euchre. It was always an inclusive crowd that included men, women, technicians, animal handlers, administrative staff, students, faculty, and even occasionally Harlow himself. When he was present, the decibel level may have lowered some and the language improved, but not much else changed. Except for the political arguments, the interactions tended to be playful and full of good humor. The lab softball team on which I was invited to play that summer was made up of people in every kind of role: shop personnel, staff, students, and faculty. While I do not recall the won-and-lost record from that summer, the feel of the after-game gatherings remains clear. The laughs and jokes were as plentiful as the amount of beer and food consumed. The gatherings tended to go on quite late even though the games were on weekday nights. Many of the Harlow lab people liked being around one another despite the political disagreements and the varying career goals and ambitions of staff, faculty, and students. Though the analogy to the sense of "family" has been overused, it was appropriate here.

By the end of my stay that first summer, I had come to know many of the people who surrounded Harry Harlow at the Primate Lab. They were a varied and impressive group, competent and friendly. There was no doubt that the laboratory had a clear, palpable purpose: the study of monkeys as developing intellectual and social beings. It was also clear that monkeys were seen as appropriate surrogates for exploring relevant human psychological topics ranging from normal to abnormal development. No one studied monkeys in order to understand monkeys; the monkeys were seen as essential tools to be used in order to understand humans, period.

From what I saw, Harlow's graduate students lived a charmed life. They were free to study whatever topic they wished as long as it addressed the general lab focus. No one felt that that he or she had any obligation whatsoever to work on Harlow's own research projects. In the freedom of that pursuit, the resources that were available to them were simply unbelievable. Besides monkeys, space, and equipment, the shops could fabricate virtually any apparatus that was desired. Experienced professional monkey testers were available to collect the data, and a data-entry and analysis unit could summarize and digitize the data and conduct the necessary statistical tests. I thought that if a graduate student could not be successful and productive in this atmosphere, he or she couldn't do it anywhere.

Within these constraints, there was an emphasis on doing things right, as "right" was understood at the time and in that place: clean, hygienic cages, nutritious food, competent veterinary and animal care support, trained surgeons and testers, reliable and valid test equipment and methods. Any discussion of ethics and morality was left for the political issues of the times, and with respect to research only if someone had failed to follow an established lab rule like failing to feed a monkey on time or neglecting to have the veterinarian check a bite wound or illness. That summer I never heard any question raised about having the moral authority to take over the lives of these intellectually and emotionally sophisticated animals for our experimental purposes.

As I drove back to Texas to begin another semester, I felt like I was returning from the other side of the moon. Each time in the past when I had driven back to Tech I experienced exhilaration when I saw the expanse of the west Texas high plains in front of me. This time I felt reluctance and anxiety. I had been at a psychological research mecca for over two months, and I wondered how I was going to readjust to the limited opportunities and facilities that were available to me at Texas Tech. This time I was disappointed to be back.

*   *   *

During the early part of the 1967–68 academic year at Texas Tech, reflections on my summer experience at Harry Harlow's lab continued to sift into my daily thoughts just as the west Texas wind blows fine red sand into every crevice of even a well-sealed house. The reverie was nurtured by frequent inquiries from both fellow graduate students and faculty interested to know what new things I had learned about developmental theory, the Wisconsin environment, and Harlow the man. Among other things, I told them that the experience had introduced me to research with rhesus monkeys who, by their very nature and complexity, challenged the simplified theories of animal behavior and seemed to offer opportunities to validly model complex human behavioral problems without the troubling ethical restrictions inherent in much human research. I wanted to be a part of the research world at Wisconsin, but I knew that this was not likely to happen. I had to make things work at Tech.

My advisor at Texas Tech was a good and serious man named Sam L. Campbell. As teachers and researchers, Sam and several other faculty represented the radical behavioral dimension of psychology in a rather traditional department of clinicians, physiological psychologists, and personality theorists. His radical behaviorist credentials were strong, as he had earned his PhD under the mentorship of Skinner himself. He was proud of the fact that he and Skinner had designed an apparatus, the "shock scrambler," that eliminated any possibility that a rat subject in an experiment where it was intended that it receive an electric shock from a grid floor might find a way of escaping it.

As a teacher Sam was challenging, strong minded, and argumentative. A central part of his graduate training approach in radical behaviorism was his demand that we examine the language we used when describing behavior. He steadfastly required that our descriptions remain limited to what could be seen objectively in any given interaction and did not stray into any "mentalistic" unobservables. As Skinner put it in 1968: "To argue that an animal does something because it finds satisfaction in doing this, or because it intends to get something out of this; to argue that people follow customs because of what they feel or because of their attitudes toward each other; to argue that a kinship system gives us an insight into the savage mind— these are mentalistic explanations which were wisely rejected."[8] To get a sense of the extent of this linguistic restriction, consider how Sam described the disadvantages of the dog-restraining device called the Pavlov sling: "We found that the Pavlovian leg hoops and other applicable techniques were ineffective against vocalizations and extremely vigorous and topographically varied muscular actions of 2-year-old beagles." Most people would have simply written that the dogs "barked and struggled to escape" while in the Pavlov apparatus. But to radical behaviorists the word *struggle* indicated an inferred internal feeling state that was not directly observable and therefore needed to be excluded from a behavioral analysis.[9]

While Sam considered the "mentalistic" psychology of Harlow and his colleagues to be wrongheaded, he had supported my summer internship very strongly. He was certain that I would return recognizing the superiority of the Skinnerian methodology and philosophy. He was correct in that I questioned the language the researchers at Wisconsin used to describe monkey behavior because it assumed the existence of a wide variety of unobserved states like hunger, curiosity, anger, anxiety, thought, and even love. It is easy for me to see now how this stand for so-called exact scientific language also helped to displace and steer us away from acknowledging the distress and suffering experienced by the experimental animals, but at the time it was a very engaging viewpoint because of its consistency with the scientific value of objectivity.

At the first lab meeting of the semester, Sam suggested, to the surprise of all of his students, that we change course and devote the lab's mission to the study of learning in insects. He explained that exploring insect learning ability from a comparative psychology perspective would be quite fascinating and that studying insects would force us to keep our analysis of behavior to the observable level, as it would surely minimize the tendency to explain the behavior of our subjects by referring to their "thoughts," "desires," and "intentions."

As a result of this shift, I spent the year working with harvester ants, helping to devise a Skinner box that would work with these small arthropods and collecting data showing that they were actually capable of learning to depress a lever in order to receive a drop of water. Watching these ants with the aid of a large lighted magnifying glass, I could see their eyes and mouthparts and the hairs that poked out here and there on their bodies. I began to think that if I tried hard I could learn to identify individuals through their unique characteristics. And when I found an ant that had become an experimental subject dead, I felt a twinge not fully accounted for by the waste of my personal investment.

I brought this feeling up with Sam at one of our regular meetings, and for a moment I thought he understood what I was saying as he nodded knowingly. Then, after realizing what was at stake psychologically, he narrowed his gaze and asked me to analyze my verbal behavior in terms the reinforcing contingencies. Any serious discussion of the struggle of thirsty harvester ants in our plastic test chamber and my personal reaction to it was set aside. I felt a little foolish for having brought up the matter.

I was truly engaged in this work, but doing research with insects wasn't my preferred path. What I really wanted to do was to study the behavior of animals at the other end of the evolutionary spectrum.

In the spring I took a job as one of the animal facility caretakers in order to avoid asking my parents for another loan. Since I was spending so much time at the lab anyway, I reasoned that I could do the job in my spare time and during normal work breaks. As luck would have it, during my first week on the job the lab students completed their rat experiments and about sixty "used" rats became my immediate responsibility. My supervisor, Fred, told me that we needed to euthanize the rats immediately so as to make room for the next batch, which was ready to be moved from the breeding colony into single cages.

This proved to be rather difficult. First, as I examined the rats targeted for death, a female student entered the room and, after an awkward moment, pointed to one of the rats and said, "That's Enid. She was my rat in the Psychology 101 lab.

"I came to say goodbye," she continued. "I learned so much in the class thanks to her."

I didn't know what to say. She went on to describe what a great rat Enid was—so smart and so gentle. She told tell me how few errors she made during the maze training and that she was the first rat in her lab section to learn to press the lever for food in the Skinner box.

Only then did I realize what she was doing: pleading with me to spare the life of her rat. I asked her if she wanted to take the rat home with her. She said yes, and so I put Enid into her canvas bag along with some rat food and sent her home, securing a promise that she would tell no one about the transaction.

Afterward I stood in the empty rat room thinking about the woman's courage and motivation. In Enid she saw an intelligent being capable of responding reciprocally to gentle human treatment. She saw cleverness, curiosity, and intention in the way the rat negotiated its life. As a result she felt she owed the animal a chance at a continued life.

I, on the other hand, was deep into the process of learning how to reject the existence of these so-called mental attributes by focusing on cold depictions of observable behavior. I was aware that in so doing I removed myself from having to reflect on animals' other ethically problematic internal states, like suffering, anguish, distress, and torment, which would claw at the conscience and raise questions about the justification for harming.

My conscience was troubled nevertheless as I set about completing the task at hand. I placed about ten rats into a black plastic bag, grabbed the bottle of chloroform, and exited the building into the slightly chilly Texas night. Opening the top of the bag, I quickly poured the liquid directly down onto the mass of black-and-white bodies at the bottom, then twisted the top closed. Instantly the rats began to squirm in an absolute frenzy.

After about three to four minutes, the bag became silent. I walked back into the building and called the elevator. As I waited for it to arrive, I noticed that my arms and hands were getting warm from the body heat emanating from the bottom of the bag. It somehow seemed wrong that the light chill of the night to which I had been exposed should be warmed by the bodies of the dead rats that I had just brutally suffocated. As I dropped the bag into the dumpster, I felt mostly relief, but there was a dull edge of sadness in my throat. I remember wishing that the rats had not been so easy to handle, almost trusting. At least Enid had escaped.

When I showed up for work the next afternoon, Fred informed me that I had not succeeded in killing the rats. When sanitation workers had come to empty the dumpster, they'd found the rats running around inside it. Since I had used up the last of the chloroform, we took the remaining rats out onto the roof where my failed euthanasia episode had taken place the night before, so that he could show me an alternative method of killing them. He took a large male rat in his hand, turned to face the brick parapet wall that followed the edge of the building, reared back, and threw the rat at the wall like a baseball pitcher throwing a fastball. The rat made a pop as it hit the wall, fell straight down onto the gravel-covered roof, quivered, and then lay totally still in the shadow of the wall.

Fred wanted me to "stun" the remaining rats in this way but, sensing my reluctance, suggested that he do the throwing while I assisted by handing him the live rats and picking up the dead ones. He warned me, however, that if I was going to continue as an animal researcher, I would have to get used to a certain amount of gore. It came with the territory.

For the next hour rats were slammed into the brick wall, dying instantly or soon thereafter. When we were done, I carted off five bags of bodies to the dumpster. As I washed the cages, the residual feeling was strangely less intense than it had been the day before when I had tried to kill the rats with chloroform. With that chemical I could hear the struggle, the running, the clawing, the futile attempts to escape the lung-searing gas. There was no struggle with the stunning approach. In the end it felt as if I had participated in something like a football conditioning drill that was intended to make me stronger and a better competitor. I actually felt like I had made an advance.

Academically, the year was going very well. My core classes in experimental design and statistics were both difficult and interesting. I also felt that I was making headway absorbing the radical behavioristic philosophy of my mentor. I was finding that I could hold my own in classes and late-night debates, especially with the clinical students. I challenged their psychoanalytic jargon and concepts and asked how they knew that the unobservable psychic conflicts that they used to explain behavior were actually taking place. I was also influenced by another Tech faculty member, Henry Cross, who in class lectures likened clinical psychologists to TV repairmen who, lacking any basic knowledge of electronics, were reduced to banging on the chassis of a broken set with a hammer and hoping for the best.

I was becoming a true believer in the reinforcement approach of behaviorism while rejecting the psychodynamic thinking that had originally brought me to psychology. As I became increasingly inured to the suffering and deaths of animals in the name of education and research, behaviorism provided a theoretical framework that actively denied the existence of that which could be the basis for admitting animals into the moral community—emotional life, self-consciousness, and intentionality. My intellect was thus equipped with what it needed to easily beat back any rebellion my heart was still capable of mustering.

At the beginning of May, I received a letter from the University of Wisconsin. Since it came from the Primate Lab, I expected that it was a note from one of the people I had befriended during my internship. To my utter surprise, the letter was from Jim Polidora, the man in whose lab I had worked but whom I had never met the previous summer. The letter was short and to the point. He offered to sponsor me as a graduate student in the Department of Psychology and provide financial support by way of a research assistantship. While I would be expected to conduct grant-supported research in learning and psychopharmacology at the Primate Lab, I would have ample opportunity to pursue my own interests as long as they mapped onto the themes and purpose of the lab. He gave me several weeks to decide.

I was elated beyond belief. Here was my ticket to the world of primate research—and not just to that world but to its elite capital, where the support for graduate research was extravagant and the possibilities seemingly limitless.

2

Induction

If the monkeys are running the next level of existence, I am in trouble.

JONATHAN K. LEWIS

I knew at once that I would accept Polidora's offer to go to Wisconsin, but issues of transition came immediately to mind and leached out some of the joy that I was feeling. How would Sam Campbell react to my leaving his lab to go to a lab headed by Harry Harlow, a man whose science he believed was experimentally wrongheaded? Did I want to give up the relative certainty of my place at Texas Tech for the uncertainty of a new program and work under a man I did not really know at all? Could I adequately compete with the other graduate students at the department and lab, who all seemed to have unmarred stellar educational backgrounds quite different from my own checkered past? Finally, how would my dear girlfriend Kim, a young woman with strong family ties to the Lubbock area, take the news? We were pursuing a relationship with definite serious undertones, but no real declarations had been made by either of us. Would she want to go with me?

Over the next couple of weeks the answers to the nonintellectual questions began to work themselves out. Sam Campbell smiled broadly when I told him about the offer. There were no petty feelings of rejection on his part, only wishes for my success. Most everyone I spoke to, students and faculty alike, said that the opportunity was too good not to accept. Kim affirmed that she wanted to go with me to Wisconsin but was certain that her parents would not permit it unless we got married, and she was reluctant to disrespect their wishes. Marriage had not been in the front of my mind, but I knew I did not want to leave without her.

A quickly arranged wedding took place two months later, with the reception at Kim's parents' large and beautifully appointed home. My parents and sister came down from New York to participate in the ceremony; my father was my best man and my sister a brides-maid. While I had the distinct impression that Kim's parents did not care much for me or my working-class family, they were cordial and welcoming as required by Texas social norms. I know that I had failed to communicate to them the value of primate research and the career opportunities that would follow graduate education. I also suspect that some of their reluctance was rooted in their concern over the at times violent antiwar demonstrations that had taken place on the Wisconsin campus the previous fall. Kim's father was a World War II veteran and deeply offended by the student behavior he saw in the news.[1]

The goodbyes to my close friends and teachers at Texas Tech were carried out with a haste that did not reflect their importance to me. But I felt an incredible urgency to get to Wisconsin. It was as though I had to get there before Polidora and Harlow had a chance to change their minds about my admission. Kim and I left all our guests stand-ing on the lawn as we drove off in my 1966 Ford Mustang, packed to the roof with books and clothes and towing a small U-Haul trailer with the few sticks of furniture we had managed to acquire. We did not drive all night, as I had the previous summer when I went by myself, but rather took two days to make the trip. The additional travel time amounted to the only honeymoon we were ever to have. There was no time to waste.

I would never regret my decision to go to the University of Wis-consin, as the experience was intellectually stimulating beyond my imaginings. Faculty members supported my development as a pri-mate researcher, and the freedom I had to explore the questions that intrigued me most was exhilarating. I was readily accepted as a full member of a research team that saw itself as being at the forefront of investigations that would unlock secrets necessary for the betterment of human society. Being a part of this team would open up opportunities that I would have missed had I completed my graduate training anywhere else. In retrospect, however, I can see that it was these very opportunities and elite-status virtues that nailed shut the lid of the box into which I had tucked my ethical conscience and sensitivity toward animal suffering. Everything around me in the Wisconsin environment confirmed that I was engaged in an enterprise of unassailable importance that required ignoring evidence that the lab animals suffered needlessly as a result.

*　*　*

Our first day in Madison was spent looking for an apartment. Because of my impatience, we quickly rented an apartment in the small farming community of McFarland, about thirty miles north of Madison. (The decision was made without examining the apartment very carefully, which turned out to be a serious error in this land of subzero winters. Though in the remaining weeks of warm weather we feasted each evening on fresh sweet corn sold by the side of the road near our apartment, we suffered the next winter as ice cracked the roof and made it leak, and the furnace stopped working during the coldest nights.)

Leaving Kim to do most of the homemaking on her own once the U-Haul was unpacked, I drove off to the Primate Lab to focus on research. In so doing I consciously instantiated a principle that research always had priority, no matter the circumstance. I assumed that Kim understood this value system and was content with her part in it, though we never discussed it directly.

When I walked through the doors of the Department of Psychology Primate Laboratory that day, I felt more certain of myself than I had the previous summer. This time there were immediate handshakes and smiles of welcome as I walked down the main hall and experienced the unique animal odors as pure perfume. I went directly to the main graduate student office in hopes of finding an open desk. The dark wood door of Room 116 framed a large glass center pane on which the names of the graduate student inhabitants were professionally painted as if they were the partners of a law firm. The sense of specialness that this small detail provided was exactly what was intended. I felt gratified when my name was added to the list of "partners" a week later.

I turned the use-polished bronze doorknob and entered. The room contained six gunmetal-gray desks arranged longitudinally in two rows of three. A new IBM Selectric typewriter sat on a narrow Formica table close to the entry door. A man seated in the furthest desk on the left—always reserved for the most senior graduate student—rose up and walked toward me with his hand extended in greeting. Monte Senko was a solid, good-looking man with a laconic quality that I had noticed the previous summer, suggesting that he bore a long-standing heavy emotional burden. But at that moment what meant the most to me was that he greeted me with friendly acceptance, like a peer.

I asked Monte how I should go about signing up for a desk in the office. He casually looked around, pointed to the last desk to his right, and said I should put my things there. With that he turned and sat down at the typewriter, inserted some paper, and began typing with the speed of a legal secretary.

I put my small bag of books on the indicated desk and sat on the chair, consciously savoring the moment. I was now a student with a home. The ease and informality illustrated by this simple episode were characteristic of the lab operation in general. There were few pretentious barriers to communication and getting things accomplished.

The following day I went in search of my new advisor, V. J. (Jim) Polidora. I was excited and nervous to finally meet him in person. I went to Polidora's lab and found the supervisor, Wayne Thompson, running monkeys in the automated learning apparatus. I was happy to see him. We sat down together and began to catch up on what was new with both of us. I told him that I had gotten married and that I was really happy to have been accepted into graduate school at Wisconsin. His response to my later statement was that he had assumed that that would happen. It was good to hear his affirmation.

I told him that I was ready to meet my new boss and asked where I could find Polidora.

His expression changed dramatically. His head cocked slightly to one side as his facial features scrunched into a wince and his jaw dropped a little. "Do you mean you haven't heard?"

"Heard what?" I responded.

He slid his chair closer to mine, looked into my face, and said, "Polidora is not coming back. He's gone. He is staying in California."

Wayne stood up and turned the chair around so that he straddled the seat with his arms extended over the back toward me. Having secured my full attention, he said that he understood that my research assistantship had been transferred to Jim Sackett. After a moment of stunned silence, I had to admit that this was even better, as Sackett had been the most impressive faculty member I had encountered during the summer internship, Harlow included. Wayne assured me that all would be fine and told me to go find Sackett and work out the details.

Standing in front of Sackett's door, I looked again at the drawing of the two copulating chimpanzees and knocked. I heard the sound of a voice and just assumed that it had said to enter. The office was long and narrow, with a desk facing the far wall and a long table with bookshelves above it on the right. The table surface was completely covered with stacks of computer printouts, and piles of books were haphazardly placed on the shelves above it. Sackett turned from the desk to face me and asked me to sit down. He was dressed informally in a short-sleeved white shirt and gray cotton pants. His face was tan, and his mouth was still outlined by the Fu Manchu mustache he had sported the previous summer. He showed only the slightest hint of a smile, which had the effect of throwing me off balance.

I began to speak right away, saying that I had just heard that Polidora was gone and that I might be working for him. I continued that I had not heard a word about this change.

Before I could explain the details of my angst, he interrupted me, saying offhandedly that it was true he hadn't called me. That was it. In other words, he did not want to hear any more about my emotional upset and I should stop talking about it.

This was an attribute that I came to appreciate about Jim. He was unsupportive of complaining for no purpose beyond general catharsis. However, if you had a real problem about your research or courses, he was always available to help.

We—that is, he—talked for about another twenty packed minutes, during which he told me that he assumed I wanted to study learning and that I ought to consider studying the effects of early experience on learning. He directed me to find some space in his lab, come back when I had some research ideas and equipment requests, and register for some courses that interested me. He told me to not forget his seminar on animal behavior. Finally, he asked if I had any other questions.

During the ensuing pause, he turned toward his desk again and began to examine one of the printouts. I turned to leave, saying that I was happy I would be working for him. He mumbled something in response that I could not understand. I asked him to repeat it and he said, "With me." Still not understanding, I asked him what he meant. "You will be working *with* me," he said a little more loudly but still not turning around.

It was later that I understood the meaning of the distinction. He was telling me that I was not one of his employees, I was a student, and he expected creativity and collegiality, not conformity.

I wandered around the lab, reacquainting myself with people and facilities. I stopped by to see Ken Schiltz; Harlow's editorial secretary Helen Lauersdorf; Chris Ripp, the colony manager and director of the nursery; and two of the animal technicians with whom I had worked the previous summer, Laverne Rossler ("Ross") and Steve Bruns.

Once I was out of the administrative offices and back in the animal rooms themselves, I immediately rediscovered how loud the animal holding rooms could get when the monkeys housed there began to screech, pound the stainless-steel walls, and shake the metal doors to their cages in a choruslike crescendo. At its peak intensity, the noise from the monkeys was reminiscent of the corrugated-paper plant at which I had worked in New York when I was a teenager. I did not wonder, as I do now, about what the monkeys intended to communicate with these loud displays. That these acts might be understood as tumults of aggression and frustration arising in part from the predicament of their laboratory living arrangements, and directed at the human perpetrators, never once occurred to me. At that time I simply saw the monkey-room noise as something I needed to get used to, just like the noise at the factory.

By the late afternoon I had worked myself up to the third floor, where Sackett had designated laboratory space. I walked cautiously into a large room containing several large housing apparatuses, each of which housed three rhesus monkey families in separate and relatively spacious wire "apartments" arranged around a central play area. The play area could be accessed only through small doors too small for the adults to pass through. In one of the units, three or four young monkeys were involved in a very exuberant play episode and took no obvious notice of my presence. The monkeys chased one another up ladders and around a spinning disk, with the roles of chaser and chased switching every few moments. Occasionally they would coalesce into a ball of bodies, wrestling and gently biting each other. The adults, sitting on perches that brought them to my eye level, took more notice of my presence, the large males demonstrating some halfhearted threats as they yawned widely and revealed their enormous daggerlike canine teeth. All the monkeys, young and adult, looked incredibly healthy. Hours of daily mutual grooming had left their brown-gray coats shiny and neat.

I found out later that this was the "nuclear family" study designed by Margaret K. Harlow, Harry Harlow's wife, and was not a part of Sackett's lab. Among other goals, the study was intended to test how adult male rhesus monkeys would interact with infants and the young if forced to share space with them on a long-term basis. Many field primatologists had misgivings about what would happen in such an environment, some fearing that the males would not tolerate close contact with infants and might injure or kill them. The study represented the psychological approach to studying primate behavior, as opposed to the ethological or evolutionary anthropological approach in which primate behavior was studied in the context in which the animals had evolved. By creating this unique but "unnatural" living environment that the monkeys would never encounter in the wild, Margaret Harlow was trying to learn something about the monkeys' behavioral capability—what was possible rather than what was normally expressed in wild living monkeys.

Over the next three years I would find myself visiting these groups frequently, mostly for the sheer joy of watching the complex social interactions. There was always much play between the young and warm attachment between the young and their adult family members. In the late afternoon the young monkeys were chased back to their family units by the technicians, and the small entry doors to the play area were locked until the next morning. At these times I could watch as each of the family groups huddled up and made ready for sleep. On one of my late-afternoon sojourns, I watched a four-month-old female suddenly crawl up her father's thigh, swing onto his chest, and tug on his lower lip with both hands until he opened his mouth wide. Then she pushed her face into his open mouth like a circus performer putting her head in the mouth of a lion, and even grabbed one of his razor-sharp canine teeth. I do not recall hearing any story about the young ever being injured by males.[2] Even in this contrived environment of enforced adult monogamy I could get a glimpse of what monkey life was like outside of the severe limitations imposed by living in individual cages and being used in behavioral and surgical experimental interventions.

Now, on that first day of my new position, I retraced my steps and found Sackett's lab. There I encountered Dick Tripp, one of Sackett's professional research associates. I explained why I was back at the lab, and he welcomed me. He thought I would enjoy working for Sackett, who would give me a lot of freedom and support. Dick then returned to his work, and at his invitation I began to explore what was around me. It was a different world from the nuclear family environment.

The research area itself was constructed of uncovered cinderblock with metal beams and electrical cables clearly visible. The fluorescent lights hung unceremoniously from the bare ceiling. This no-frills room contained a number of cream-colored sound-attenuating chambers that measured about four feet square. The chambers had heavy chrome door handles, the kind you might see on refrigerated meat lockers. I looked inside a chamber through its observation port and could see that inside lived a single infant rhesus monkey in a stainless-steel wire mesh cage. These were some of the total social isolation chambers of which I had read. The monkeys inside had been separated from their mothers at birth and purposely placed in these chambers as a way of preventing the social experience that would normally be provided by the infant's mother, siblings, and peers.

In addition to the chambers, there were several other wire cages, each housing a single young monkey. This housing arrangement was called "partial isolation rearing" because the monkeys were exposed to social stimulation from the other monkeys, who were visible to them, but did not receive physical contact. Some of the monkeys carried packs the size of a small matchbox on their backs that recorded heart-rate data. The packs were secured to their backs with amber-colored rubber tubing that looped over their shoulders. The room was full of electronic equipment as well as a cache of basic medical supplies like sutures, antibacterial solutions, ointments, and gauze pads. These supplies were used to treat the cuts, abrasions, and irritations that developed on the skin of the monkeys carrying the data recorders. Compared to the young monkeys living next door with their nuclear families, these animals looked to be stricken by the experimental manipulations they were enduring. There was no joy that I could see, only what I considered to be cutting-edge developmental science. Dick made it a point to tell me that medical care in the form of veterinary and nursing expertise was always close at hand.

I found an empty room in the far corner of the lab, just a few feet away from one of the total isolation chambers. Finding a small wooden table and chair that looked like discarded furniture in the hall, I placed it in the room as an act of territorial marking.

During the next several months I scouted around the lab for programming and test equipment. I was astounded by the variety and amount of surplus testing and electronic programming equipment that could be found stored in unlocked closets, hallways, and unoccupied rooms. Often when I inquired about the ownership of the equipment, my question was met with a shrug. In those cases I taped a note to the vacated space and carried the equipment up to my room. By the time the weather began to turn really cold, my corner of Sackett's lab was looking like a serious scientific enterprise, and no one had come to reclaim any equipment. I eventually realized that the budgets of many of the lab investigators were so flush that they could easily replace "lost" equipment, and they were always upgrading their technology and casting off the used materials. To fill the needs not met by the equipment I appropriated, Sackett approved additional expenditures of several thousands of dollars for new equipment. He never questioned my requests. I began to realize that he trusted me not to make unnecessary purchases and to create a sound research setup.

As Sackett suggested, I began reading in the area of developmental psychology and the research on the effects of early experience on behavior, and on learning in particular. So each day, after the morning lab ritual of coffee, donuts, and conversation was adjourned, I went off to the lab library, with its complete collection of books, journals, reprints, and free copy machine, to study until it was time for an afternoon class. It was immediately clear that much of the research in developmental psychology, including that which I was seeing around me, was dedicated to determining how much of animal and human behavior is accounted for by "wired-in" mechanisms bestowed by genetics and heredity and how much is acquired through experience and learning. In the course of that reading I found that unraveling the nature–nurture question at that point in history required raising newborn animals in laboratories apart from "normal" species caretaking, so that the experiences of the animals could be programmed according to the questions of the researcher. Raising animals in conditions of social or sensory isolation during hypothesized "critical periods" was considered one of the standard methodological tools for parsing the distinction between behaviors originating in nature and those that required nurture to develop. The consequences of isolation were often described graphically in the studies: birds could not sing; rats froze with anxiety when placed in novel open fields;

ducks tried to mate with humans; dogs couldn't effectively defend themselves against aggressive litter mates and repeatedly pushed their noses into lit match flames that were held in front of them, as if they didn't feel pain.[3]

In the research reports, the procedures and equipment used to create the isolation manipulations were described unflinchingly and without a hint of ambivalence. It was as though drastically altering the lives of animals had no ethical questions attached to it at all. This was the same message that I had been hearing since undergraduate school, but now I was seeing it not just in articles by recognized leaders in psychological and biological research but in designs implemented in my daily reality. Some authors, apparently believing that some justification was necessary, took the tack of reminding the reader of the obvious moral restrictions on doing similar studies with human children. They might also point out that information collected on children raised in deprived environments like orphanages was so questionable as to be useless for validly addressing the question. Producing harms in animals in the name of science was by definition an ethically neutral act as long as you considered, even in the most glancing fashion, doing the work in humans first.

It is amazing to me to think now that this so-called reasoning made unquestioned sense despite its logical absurdity. How could the ethical need to avoid causing significant harm in one group of beings automatically legitimate the causing of similar harms in another group chosen specifically because its members were capable of being similarly harmed? Despite my intellectual conformity to this "principle" at the time, I cringed when I read about the studies that isolated canine puppies for months, as I could easily recall the dejection expressed by my dogs at home when they were left alone for only a few hours.

I was learning again that my own commonsense concerns about what constituted appropriate animal treatment, grounded as they were in a vague rule of not harming the defenseless, were not a valid basis for critical argument about animal treatment in the lab. Although a general ethic of humane behavior that protected vulnerable humans in everyday life mapped onto the world of human research, an analogous ethic for animals did not necessarily apply to their scientific use. This lack of symmetry could make sense only in a world that was deeply bifurcated into the categories of human and animal. This bifurcation existed despite the fact that Darwin had shown us a hundred years earlier that there was no such separation, that there was biological continuity from animals to humans. As scientists, then, we were Darwinians when it served the purpose of arguing that animal experimentation was relevant to understanding human functioning, but when moral questions derived from that attribute were poised to challenge the justification of animal experimentation, we became staunch noncontinuity theorists. Recognition of this essential hypocrisy was nowhere to be found in either the literature, nor, as I would find, in the views of my colleagues. Similarly, there was no effort to engage in a legitimate ethical analysis or grapple with the question hanging invisibly over the whole lab building: *what are the morally relevant differences between humans and animals?*

Fortunately for me, because of my need to learn about the findings on early experience effects in monkeys, I found that Harlow had published several articles in 1964–65 that described in detail the consequences of raising rhesus monkeys for the first three, six, or twelve months of life, or the second six months, in total social isolation. In these studies total isolation consisted of confining newborn monkeys in two-foot-square solid walled sheet-metal chambers that were continually illuminated and filled with constant white noise to block out extraneous animal and human noises. Like the other authors I read, Harlow described both the isolation environment and its consequences in graphic and nonapologetic language. He described the aggression seen between previously isolated monkeys and controls in social capability tests as "abusive" and at times so "murderous" that intervention was required to prevent life-threatening injury. He described a monkey starving to death due to the environmental shock that it experienced upon release from total isolation.

However, there were occasional departures from this pattern that I noticed. For example, in one article Harlow acknowledged that monkeys in total social isolation were being subjected to "extreme"[4] conditions and, perhaps most revealing, that they were serving an experimental "sentence."[5] Was he acknowledging that isolation was akin to a drastic punishment meted out to an innocent animal in the name of science? No one I asked about these words thought anything about it other than to say it was an example of Harlow's flair for the dramatic and his ability to create a picture in the minds of readers that illustrated what was happening experimentally.

The data in the papers I read showed that monkeys raised in isolation for the first six or twelve months of life emerged from confinement with what would become known as "the isolation syndrome." It was considered a syndrome because the effects of the isolation on individual animals were consistent and predictable. The syndrome comprised high levels of fear, deficient social behavior, reduced environmental exploration and curiosity, increased self-directed behavior like huddling and self-clasping, and bizarre stereotyped behaviors such as repetitive pacing and flipping and even self-injurious behavior that sometimes left open wounds on arms and legs. Later, as more mature animals, monkeys raised in isolation were found to have disordered sexual posturing and incompetent maternal behavior.

On the other hand, monkeys reared for only the first three months or the second six months of life recovered some amount of social competence if provided social contact with nonisolated controls. There appeared to be a critical period between six and twelve months of age during which social stimulation was necessary for normal development of the parts of the brain responsible for social behavior; if social stimulation did not occur during this critical period, the monkey was left unable to recover the lost capabilities despite attempts at rehabilitation. However, when the socially devastated monkeys were tested with a standard battery of learning problems, they were found to be indistinguishable from a comparison group of controls. In all the research I read, the controls were monkeys who had been raised alone in open-mesh wire cages; while they could see, hear, and smell other monkeys, they were not provided the opportunity to interact with other monkeys except at a distance and never with physical touch. I wondered if such monkeys could really be considered to represent the "normal" unmanipulated condition.

Harlow summarized the findings of these experiments by stating that isolation disabled the "social mind" to a greater extent than the "intellectual mind." He pointed out that the failure to find intellectual differences between the isolated and cage-reared monkeys was at odds with the human findings, which were based on nonexperimental data from children raised in orphanages and other deprived environments. I took note that in commenting on this discrepancy, Harlow made a crucial and revealing interpretational turn. Instead of entertaining the possibility that the failure to replicate the human findings indicated that rhesus monkeys were not good models of human intellectual development, he argued that the human data must be erroneous. This conclusion demonstrated Harlow's confidence in animal models and their relationship to human behavior; if the data on humans and on monkeys are in conflict, the problem is most likely with the human data. This interpretation had a powerful impact on me, coming as it did from one of the most important experimental psychologists of the time.

At the time, this incredible position did not raise my skepticism but instead increased my sense of confidence in how I was starting to spend my research life. It also helped to dispel any misgivings I had about the costs the animals were paying for the sake of science. Several years later I would firmly align myself with this position in an article I coauthored with Harlow and Stephen Suomi that appeared in the *American Psychologist.* "Sometimes when monkey data fail to generalize to human data," we pronounced, "the answer lies in the superiority of the monkey data and the need to revise those data that are human."[6] I did not seriously consider that such an interpretation might also be a self-serving maneuver that insulated the animal work from criticism.

Data on the elements of the isolation syndrome had an obvious relationship to issues related to the nature-vs.-nurture debate, but Harlow believed they might also be relevant to the development of animal models of human neuroses and psychoses. His suggestion was initially tentative because "subhuman animals probably do not possess many complex personality mechanisms that suffer distortion before or during psychiatric breakdown." However, he put forward the possibility that rearing monkeys in isolation might well be able to model important aspects of human infantile disorders that result from deficient parental care during the period when a child's personality structure is not yet fully developed. In staking out such ground, he was aligning this dimension of his research to the influential work on experimental neuroses that had begun with the Nobel Prize–winning physiologist Ivan Pavlov in the 1920s, which Harlow described in the mid-1960s as only "possibly rewarding."[7] During the next several years Harlow's skepticism about using monkeys to model adult human psychiatric conditions would dissipate, however, and his use of procedures even more drastic than total isolation would bring both acclaim and condemnation.

For me, this clinical connection to primate research added greatly to its desirability and importance. It is one thing to put a baby monkey alone in a two-foot-square cage for a year as a way to determine whether the ability to learn was wired in or required complex environmental input during development, and quite another if the disabilities that emerged as a result of the isolation were related to human psychopathology and suffering. In the first case you more or less knew what the outcome would be but needed to demonstrate it empirically in order to validly establish it as a scientific fact. In the second case you were creating a prototype that would make it possible to investigate the causes of human suffering and possibly find effective therapies.

After completing my literature review, I was convinced that there were many approaches to testing the learning capacity of isolate monkeys that were worth trying. I decided to apply what I learned about operant strategies with Sam Campbell at Texas Tech. In this approach, developed by B. F. Skinner, animals are deprived of food or water and then trained to get access to a food or water reward by operating a simple switch like a telegraph key (I had been successful in training harvester ants to depress a tiny lever to get access to water). Once the animal learned the relationship between the action and the reward, one could make the reward contingent on a variety of different response patterns. One could, for example, increase the number of responses necessary to get a reward, or require the animal to withhold its response for some prescribed period. In this way I could test whether isolate monkeys were different from controls in the speed at which they learned to adapt to shifting cognitive demands or in the approaches they took to learning. An additional benefit was that all the different tests could take place in the same piece of apparatus, reducing the need for the isolate monkeys to adapt to different circumstances—something they had difficulty doing. I saw no evidence that this type of testing had yet been tried in assessing the effect of early experience. I met with Sackett to discuss my experimental plans. He listened carefully and asked several practical questions. He approved of the approach, and before I left his office I had a list of eighteen monkeys that I could use in my research.

The rhesus monkeys who were placed under my research supervision included six who had been reared in total social isolation for the first nine months of life, six raised in open wire-cage partial isolation, and six who had been reared with their mothers and with access to peers for the first year of life, after which they had been transferred to individual caging. Each group of six was composed of three males and three females. After getting their locations, I started to visit them each day, trying to get to know them and prepare myself for dealing with their idiosyncrasies and anything that might make testing difficult. At first I just stood by their cages and fed them raisins; then I provided them their daily ration of biscuits and fruit; finally I began transporting them to the scales so that I could track their weight. It didn't take long before I could recognize their faces and identify some of their unique habits. They were all very different animals, both in appearance and in behavior.

E25 was a female who had been raised in partial isolation. She had very pale skin and a coarse, curly, rust-colored coat. She was always very cooperative, never balking at the entrance to the transport cage when I went to retrieve her. She moved her slender twenty-plus-pound body calmly and almost daintily. When I dropped her daily biscuits into her feeding box in the late afternoon, she collected them in a very particular way. Instead of pushing a number of the biscuits into her cheek pouches as most other monkeys did, she would take them one at a time out of the feeding box, rub the loose crumbs off them, and line them up end to end around the inside edge of her cage where the walls met the floor. She fussed with their presentation, as if she had a specific design template in mind, until they were all in place. Then she would sit in the center of her cage, surrounded by the biscuits, and eat them one at a time. I assumed that she had developed this behavior as a way of stimulating herself and reducing slightly the monotony of her living environment.

G49 was another female, but one who had been raised by her natural mother and with access to peers of both genders during the first year of her life. She was as close to a "normal" monkey as could be produced in the laboratory. She weighed about twenty-two pounds and had a very thick gray coat that made her look like she was prepared for the cold of a Midwest winter. Between her eyebrows was a small nubbin of pink-colored flesh that became a darker red each time she reached the middle of her menstrual cycle. While I was getting her accustomed to my presence and the daily testing regimen, she directed many facial expressions and body postures at me. She lip-smacked with her lips parallel, and at other times with her lips pursed and turned upward. She would circle to show me her butt or extend her chest against the front of her cage in an invitation to groom her. She, too, was cooperative but seemed to have a wider range of emotional responses available to her than E25. Sometimes she was just not ready to go, and other times she acted like I had annoyed her and hooted out a threat to me as she entered the transport cage.

A fellow graduate student, Charley Pratt, had tested G49 for her social competence for his master's degree. He told me that she was one of his favorite monkeys because she never attacked another and would huddle next to the individuals who were frightened by the testing environment as if to comfort them. This description of G49 as a kind and compassionate monkey violated my behavioristic perspective. I interpreted her "compassionate" activities not as a personality trait or choice but as the result of the contingencies of reward to which she had been exposed during development. Charley resisted my interpretation, insisting that I was selling the complexity of monkey behavior short. Since he was by far the brightest student in the lab, I could not discount the meaning of his observations entirely.

E12 was a large, handsome isolate male who weighed about thirty-five pounds. He had a sparse light-gray coat with light-brown and yellowish highlights. While his broad back, thick thighs, wide face, and powerful neck and jaws made him look like he spent his life climbing tall trees and lifting heavy burdens, he rarely even paced around his home cage. Instead he tended to stay curled forward with his face pointing toward the floor. While he was in that position, sometimes one of his legs would slowly rise off the cage floor, abduct away from his hip, and go behind his head, almost as if the leg were a separate entity with travel plans of its own. Then the leg would appear to sneak up and attack the back of his head by grabbing an ear or the flesh on the back of his neck. In response, E12 would defend himself from the attacking leg. Holding it with his hands and pulling it forward, he would bite the foot as he tumbled forward and onto his side. Once the struggle was over, he would return to the initial huddled posture and masturbate. Trying to get him to take treats from me and to shuttle into a transport cage for weighing and testing was mostly a matter of getting his attention and helping to unfold him from his curled-up, self-directed life.

E15 was a female total isolate who made many of the same "phantom limb" moves seen with E12. However, instead of defending herself from an attacking appendage, she seemed either confused by the appearance of the moving arm or leg or just frightened by it, looking as if she were trying to hide from its presence. Her interactions with me, unlike those of E12, were mostly limited to expressions of fear, which I carefully referred to at the time as avoidance. Oddly, she spent a lot of time standing up in a bipedal posture, from which she would spin in small counterclockwise circles for reasons unknown to me.

By the end of the summer of 1969 I had built my operant learning apparatus and habituated all the monkeys to the apparatus and the testing schedule, and had begun my research into how isolate monkeys adjusted to changing work demands in order to get food rewards. Things were going beautifully. I was very pleased with myself.

* * *

Jim Sackett held regular lab meetings in the large, light-filled conference room across the street from the lab on the top floor of the Wisconsin Regional Primate Research Center. At these meetings we discussed current research articles and presented our plans for experiments and the results of those we had conducted. The meetings were serious in their intent but good-humored in affect.

It was here that I began to understand Sackett's overall research plan and how raising monkeys in various forms of isolation fit into that plan. He put it clearly one day, saying that someday he hoped to be able to raise a monkey in total social isolation while at the same time providing an array of specific stimulus inputs that would render the animal socially capable upon release. I now understood why the infant monkeys in those cream-colored chambers received targeted brain stimulation and were flashed pictures of monkey faces and social situations. He was trying to kick-start social development without providing direct social experience to those infant monkeys. He wanted to discover what elements contained in social experience were necessary for the development of normal social competence and when each element needed to be present.

I loved these lab meetings and looked forward to the discussions and camaraderie. Sackett always seemed to have a new dataset that challenged our statistical knowledge and interpretational skills.

Discussion or controversy about the ethical use of primates did not generally come up during Sackett's lab meetings. The only such discussion that I can recall was started by a student named Paul Kottler, who was not one of Sackett's graduate students but found it beneficial to attend the lab meetings from time to time, and Jim made him welcome. Paul was studying the biochemistry of learning and memory in rats. At the time there was great interest in the relationship between RNA and learning, and Paul was deeply involved in this area of research.

During one lab meeting that Paul attended we were discussing newly published data from Alan and Beatrice Gardner of the University of Nevada–Reno on the acquisition of sign language by the chimpanzee Washoe.[8] Most of our discussion revolved around whether the data presented by the Gardners was in fact a compelling example of true language. Some people felt that the data suggested nothing more than stimulus-response learning, while others thought that the gestural locutions were genuine evidence of linguistic expression.

Then, out of the blue, Paul chimed in and asked, "What would the Gardeners have to do if one day Washoe signed to them 'I don't want to participate in this experiment anymore'?"

I recall this moment quite vividly. Paul had recognized that there was a relationship between the cognitive capabilities that the Gardeners were arguing Washoe possessed and the ethical implications of those abilities. The collective reaction of the group was to deflect his inquiry back to the issue of whether gestural signs were a valid linguistic representation of animals' internal preferences. Calling forth my strict behavioristic beliefs, I questioned whether animals had any subjective life that included preferences about their current or future circumstances in the first place.

With a forcefulness that he usually reserved for the lunchtime bridge table, Paul insisted that we respond to his question directly. The room got silent again. But the silence was more about whether Paul was embarrassing himself than about the ethical question he had courageously raised.

In retrospect, Paul's question was a brilliant one, but none of us, as far as I could tell, was ready to take on the profound ethical dilemma that it represented. Actually, it went beyond that. I know I didn't appreciate that it was a question that needed to be attended to in the first place, and I don't think that I was alone.

Paul's own research involved training rats to learn a particular response, usually negotiating a maze of some sort, after which their heads were guillotined off and the brain ground up and studied for

evidence of changes in RNA that might represent the new information that they had learned. The other graduate students and I often
teased Paul as he sat quietly at his desk in Room 116, asking whether
he was taking a "kill break." I wonder now whether killing rats all day,
as Paul sometimes did, prepared him to ask the question he posed.
Several times I watched Paul guillotine mice in his basement lab and
saw many mice resist as he angled their heads into the slot beneath
the cutting blade. In those cases, I thought I saw Paul hesitate and
then push past their resistance for the sake of research.

*   *   *

My first sustained and regular interaction with Harry Harlow came
in the spring of 1969, when I participated in the research seminar he
held called Mini-Experiments and Micro-Methodology. The seminar
was about designing meaningful experiments with small numbers
of primate subjects—a limitation typical in the field.[9] In this course
Harlow first reviewed the history of important psychological experiments that were accomplished with small numbers of experimental
subjects. His examples included the work of Herman Ebbinghaus,
who studied the mechanisms of rote memory in humans using only
himself as a subject; Wolfgang Kohler's work on insight learning,
completed with only a few chimpanzee subjects; Piaget's monu-
mental cognitive development theory, based on observations of his
own children; Skinner's studies of operant learning, which were at
times limited to a single subject; and his own "small-N" mother-
surrogate attachment studies. The point was that when working with
primates one did not have the luxury to make a large investment in
animals to test an experimental idea. Even more important, because
primate research must often necessarily proceed with small numbers of subjects, the researcher is faced with the problem that the
sample sizes (N) are far smaller than what statistical models say are
required for making valid inferences. Harlow's recommendation for
transcending this problem was to design experiments in which the
impact on the variables under study was apt to be dramatic rather
than subtle. To illustrate this point, he suggested that in designing

monkey experiments we should "think photographically" rather than statistically.

Harlow passed out schematic plans for several studies that made use of only two to four monkeys. The study that sticks in my mind was titled "Social Salvation through Sibling Succorance." The title was typical Harlow in its use of alliteration and poetic style. The study was an attempt to reverse the effects of total social isolation by exposing six-month-old monkeys, just emerged from isolation, to ninety-day-old monkeys that were being raised with extensive peer experience. Would the younger monkeys cling to the older isolates and attempt to initiate rudimentary play interactions with them? Would the isolates respond, thereby breaking the hold of social fear and displacing the stereotyped behaviors characteristically produced by isolation? It was a brilliant design made possible by his intimate knowledge of normal monkey development and the devastation caused by isolation. It was obvious from Harlow's description of the design that if you conducted the study, it would produce poignant images of frightened isolate monkeys and clinging and playful "therapists." (The study was in fact conducted and was successful in showing that substantial rehabilitation of the isolate syndrome was possible. One understood the results from looking at the photographs of young "therapist" monkeys hugging isolates tightly until they gave up trying to get away and then of awkward but mutual play developing between the two; there was no need to read the statistical graphs.)[10]

Given Harlow's lead, members of the class proposed various kinds of studies as the semester proceeded. Joyce Rosevear, for example, proposed to study the impact on infant development of causing mothers to reject their infants much earlier than normal by hormonally manipulating the mother's reproductive status. In this study, the image of the impact would show an infant striving desperately to maintain contact and a mother having to be increasingly punitive to create separation. Stephen Suomi, who went on to develop great stature in the field of primate development and the use of animal models, proposed to raise a newborn infant with only an adult male for a constant social partner. In a proposal titled "See No Evil, Hear

No Evil," Allyn Deets described a "sadistic little project" in which he would blind or deafen two newborn monkeys and study the effects of this sensory limitation on everyday mother–infant interactions and the infants' response to periods of maternal separation.

In my own presentation I suggested using food reinforcement to train adult socially avoidant isolate monkeys to tolerate the proximity of another monkey. Isolation-raised and socially experienced adults would be deprived of food for twenty-three hours and then trained to press a key to get food rewards inside a long narrow chamber. One isolate and one normal monkey would then be placed together in one chamber with one of the food-reward keys at each end. Over a period of days the devices would be moved closer together as the hungry monkeys continued to gain access to food. The hope was that once the isolate monkeys learned to tolerate the physical proximity of the other monkeys, some forms of social interaction would develop, and this result could be clearly illustrated by a photograph. (When I later tried this procedure, the normal monkeys frequently attacked the isolate monkeys. My attempt at creating a behavior therapy environment for isolation recovery was a complete failure.)

While very few of the studies proposed were actually conducted, I can recall no objection being expressed by anyone about the feasibility or appropriateness of any of the more invasive experiments. Even the blinding and deafening of infant monkeys proposed by Allyn Deets already had precedence in the work of the primatologist Gershon Berkson, who made a career out of studying the effect of disability in humans and experimental animal models.[11] I believe however, that many of the proposed studies captured Harlow's recommendation to "think photographically."

*   *   *

Although the ethics of doing research on monkeys was not explicitly on the agenda at the lab, the ethics of the Vietnam War held center stage. Because the local draft board was across the street from the Primate Lab, we had a ringside seat to some of the campus conflict.

It was not unusual for vehicles filled with National Guard troops to be parked along the street in preparation for potential protests. Normally there were almost daily arguments in the break room among students, faculty, and staff members over the purpose and justification of the war. The bombing of Cambodia, the Kent State killing of four protesting students in 1970, the bombing of the University of Wisconsin Math Research Center, and the start of the American troop withdrawal all provided highly flammable issues.

As I had seen during the summer of 1967, the lab arguments were intense, with much fist pounding and many raised voices at lunch and during coffee breaks. Back then the factor that had been remarkable to me was that the people who fought each other so passionately over the war worked so well with one another when research was the focus. This time I could see that this ability to hang together after bitter political argument was grounded in part in a sense of mutual respect but also in the certainty that the research taking place at the lab was in a way far more important than even the war. The war was a relatively short-term issue, while the lab research was about elucidating relationships that, when applied, might produce lasting positive changes in the human condition. Indeed I cannot recall a single prowar staff person, faculty member, or graduate student who made the decision to enlist in the military. To be fair, some members of the staff and faculty were beyond military age, and a few had already served earlier in their lives. But this was not true of all of them, and certainly not of any of the prowar graduate students that I knew. For them, as well for the rest of us who were stridently antiwar, the conflict was the business of others, not us. Our place was in the laboratory and not on a Southeast Asian battlefield.

My draft notice arrived early in the fall semester of 1969, forwarded to me by my mother. The letter was postmarked Flushing, New York, the location of the draft board where I had registered when I had turned eighteen. The letter simply directed me to report for my preinduction physical examination at Whitehall Street in Manhattan at a date I cannot now recall. I was strangely surprised by this letter's arrival. While I'd known in some vague way that it would be coming, I had entertained this odd fantasy that my research at the

University of Wisconsin Primate Lab was sufficiently important to block the military bureaucracy.

I quickly went to Room 116 to use the communal phone. Before calling, I waited until the room was empty so that I would not have to reveal my anxiety in the presence of other graduate students. I explained to the man who took my call that I was a graduate student at the university and didn't see how I could report to Whitehall Street on the designated date. I hoped that merely reporting my location and current activity would be sufficient to terminate the process. Of course it did not. Instead the man said that he could easily transfer the site of the physical to the induction center in Milwaukee, Wisconsin. He let me know that I might apply for a new student deferment once the physical exam process had been completed.

The government bus that took us from Madison to Milwaukee picked us up at the draft board across the street from the Primate Lab. I took a package of articles to read on the two-hour trip. I never got to read them, however, as there was a large contingent of demonstrators on board who loudly chanted antiwar slogans all the way to Milwaukee. As we all processed through the various examination stations, I was amazed at how patient the military personnel were even as they became targets of abuse from some of the more strident students. Curses were hurled and cups of urine splashed. Yet they remained quiet and focused on their tasks. I was ashamed of many of my antiwar comrades that day. At the end of all the examinations, it was clear that my health was good and I was a prime candidate for induction.

When I returned to the lab the following day, I went directly to Harlow's office and met with Helen Lauersdorf. I explained my situation to her, and she immediately asked for my paperwork. After a quick review, she said that she would ask Dr. Harlow to write a letter to the draft board in New York and request a new student 2-S deferment for me. She added that Harlow had sent many successful letters in the past.

Soon after this meeting Helen provided me with a draft of the letter that was to be sent on my behalf. While I don't remember the specific wording, I do recall that it gave the distinct impression that the national security of the United States and the protection of the public welfare would be better served by my being allowed to continue my research on the effects of early experience on learning in rhesus monkeys than by compelling me to become a member of the armed forces.

The letter had its desired effect, as I soon received a letter granting me that 2-S deferment. So now even the US government had given me a free pass from participating in a war in which tens of thousands of men, most about my age, would lose their lives or their mental and physical wellbeing. I felt more than just lucky. I felt somehow that I had earned this pass by committing myself to study primate models of development.

As this incident indicates, Harlow actively protected graduate students in the lab from military exposure, which is understandable given that our ability to continue pursuing our research served his own interests. His protectiveness extended to shielding us from the depredations of the Department of Psychology as well, however, and this was less rooted in self-interest than in a genuine sense of decency and compassion. The academic environment for graduate students at the Department of Psychology was notoriously harsh and at times even brutal. There were many stories circulated among the students about capable people being forced to leave the program for failing one examination or another. I remember hearing about a student whose master's degree committee judged his experimental thesis to be scientifically meritorious but failed him on his oral defense of that very thesis. As a result, he was not permitted to go on for his doctoral degree and left the program. The backstory was that this treatment was intended to send the message to the student's faculty mentor that he should leave Wisconsin and not wait for a future tenure decision, the outcome of which had already been determined.

Many students whose academic records were solid still failed one or more sections of their preliminary exams and thus suffered the same fate. It is telling that Carl R. Rogers, the internationally acclaimed expert on psychotherapy and humanistic psychology, had resigned from the department in 1963 because of what he saw as inappropriate treatment of graduate students. In his resignation memo to the psychology department faculty, dated May 8, 1963, he claimed that the department had "established a psychological climate which is experienced by most students as threatening and even punitive" and said it was "deplorable" that "only about one out of five" of the department's "carefully selected students" obtained their doctorates.

In various ways Harlow used his considerable power to protect his students from the department's irrational insistence on demonstrating "high standards" through the culling of graduate students. Like Rogers, he believed that once selected, a graduate student deserved support and encouragement, not a series of arbitrarily set hurdles.

*   *   *

Not long after I had received my master's degree in 1970, Jim Sackett arrived at the weekly lab meeting with no dataset or article for discussion. Instead he sat down at the head of the conference table, leaned forward while his right leg bounced rapidly, and explained that he had decided to leave the University of Wisconsin as soon as he was able to land an adequate position elsewhere. He said that he had already spoken to colleagues at the University of Washington Regional Primate Research Center in Seattle and it appeared that he might move there, but it was too soon to be sure. The fact that he had decided to leave Wisconsin before he had another job revealed the urgency he must have felt. He was acting like a man needing to escape a burning building.

Jim said nothing about the reason for his decision, and I don't recall if anyone asked him for one. I assumed that it had something to do with his relationship with the Harlows. There had long been vague rumors that there was conflict between him and Margaret Harlow, but I never heard him say anything that confirmed this. Then again, Jim never made derogatory or snide remarks about any faculty member or student; it was not his way.

After his announcement, Jim proceeded to go around the table and explain the contingency plan he had developed for each of us. For some of us he suggested other mentors; for others he recommended modifications of our ongoing projects that would allow us to finish the requirements for our degrees before he left. Getting another mentor was not just a matter of arranging continued access to research expertise. At Wisconsin a student without a mentor—and that mentor's financial support—had no standing and had to leave the program regardless of his or her performance up to that point.

We looked at one another in confused amazement. I recall feeling terrified. Up to that day my general plan was to stay a student for several more years, conduct more research, and take more classes. I felt that Sackett was the perfect mentor for me. He was brilliant, challenging, easy to engage in deep discussion, and open to my Skinnerian research methods. Skill-wise, he was an absolute master at statistically analyzing large, complex datasets, and I wanted to learn more of that from him. Personally, he was an unpretentious regular guy who played on the lab football and softball teams and drank with us at day's end—martinis, as I recall. I both liked and respected him.

Some days later I went to his office and sat down with him. I wanted to ask him to reconsider his plans, but his intense and resolute position at the lab meeting made it clear that was not an option, no matter the collateral costs to students. I asked him whether he had any suggestions for me about getting a new mentor. I told him that I had been thinking about John Davenport, who was the associate director of the Regional Primate Center and had a large laboratory devoted to the study of animal learning. I knew his graduate students and staff members and liked them all. Personally, Davenport was also an unpretentious person whose public presence was a mixture of self-effacement and shyness.

Jim interrupted me and said that Harlow had already agreed to take over my support. I must have looked as surprised as I felt, because he said, "He loves you." At this point he said that he thought that the move to the University of Washington would likely work out, and if so, he wanted me to go with him. I was surprised again, as I'd never been quite sure what Jim had thought of me as a student. Obviously his invitation was an illustration of his confidence.

In the ensuing weeks I met with Harlow to gauge for myself his willingness to take me as a student and provide the necessary financial support, and I found him to be very supportive. He also suggested that enlisting John Davenport to serve as a co-mentor made sense given his greater familiarity with my preferred research methodology. I met with Davenport and he agreed.

I returned to see Jim and told him that though I was incredibly ambivalent about it, I had decided to stay at Wisconsin. He said that he understood, as changing graduate schools was "a little like getting divorced and remarried." After he left Wisconsin we maintained a regular correspondence, through which his mentoring continued.

The Harlow lab was an absolute oasis from the environment of the Psychology Department. The expectations for self-direction and scientific maturation were obvious. It was only the rare and troubled student who was "given" a research project by his or her major advisor. The lab was full of teachers in addition to the professors. Members of the professional staff were available and willing to instruct any interested student in electronics, computer programming, welding, equipment construction, observational scoring methods, statistical analysis, manuscript development, methods of animal care, breeding, and the basics of veterinary medicine. The open availability of Harlow's own library, replete with reprint collection, relevant journals, book collection, and a free copy machine, was an incredible resource. Besides the information it contained, the money it saved us, and the ease of access it provided, it was a clear expression of intellectual sharing that came directly from the top. The diversity of the book collection, which ranged from anthropology to zoology, showed respect for the wider array of scientific disciplines while also making a quiet point about what it meant to be an educated person. Faculty, students, and staff members communicated openly. We worked together, played intramural softball and touch football together, and critiqued each other's research. Yes, Carl Rogers would have certainly approved of the laboratory that Harlow and his colleagues had created.

As for Harry Harlow himself, he was always open to discussion during seminars and individual mentoring sessions, and occasionally he participated in a very good-natured way in the competitive lunch-time bridge games. In his role as lab director, he was, in my view, beyond reproach. After all, he had created the original laboratory in a borrowed test room at the Vilas Park Zoo in 1930 and transformed it over the years into a facility made up of two multistoried buildings and deserving of its designation as a regional primate research center. He was the one primarily responsible for keeping the money that funded research assistantships and general research resources flowing. He was our "godfather" and protector and could activate the "old boys'" network if necessary when we were ready to graduate and seek employment. He would write beautiful and thoughtful letters of recommendation, recounting our accomplishments in the best light and describing our scientist virtues in prose that revealed how closely he was aware of us as individuals. All these factors limited the motivation for overt criticism.

What was missing in this incredible place—and in most animal laboratories at the time—was a sophisticated appreciation of the ethical obligations that arise from purposely creating harm in animals, even when it is done in the name of science. Though Harlow was a leader and innovator in so many areas of psychology, he was just a member of the crowd when it came to determining whether a particular experimental manipulation could be ethically justified. The accepted all-encompassing single ethical principle was simple: if considerations of risk and significant harm blocked the use of human subjects, using animals as experimental surrogates was automatically justified. That was all there was to it.

To be frank, I am amazed that Harlow did not apparently appreciate the ethical implications of his work. After all, he was a leader in demonstrating that rhesus monkeys were vulnerable to all kinds of harm, not just physical pain. He showed that monkeys could be emotionally destroyed when opportunities for maternal and peer attachment were withheld. He argued that affectionate relationships in monkeys were worthy of terms like *love*. In his work on learning in monkeys he vanquished the totally robotic view of the process offered by the behaviorists by offering abundant evidence that monkeys develop and evaluate hypotheses during attempts to develop a solution. Everything that Harlow learned from his research declared that monkeys are self-conscious, emotionally complex, intentional, and capable of substantial levels of suffering.

*   *   *

In the late summer of 1970, both Harry Harlow and John Davenport told me they had been contacted by Frank Logan, chair of the Psychology Department at the University of New Mexico in Albuquerque, concerning an upcoming opening for an animal learning psychologist for the 1971–72 academic year. Clearly, Logan had called to see if there was a strong candidate in their labs. The implication of their alerting me was that Harlow and Davenport considered me to have an excellent chance of landing the job.

I was as surprised as I had been the day Jim Sackett told us he was leaving Wisconsin and we would have to find new faculty mentors. As then, I was enjoying being a student and my research was going smoothly. I had developed a serious interest in the application of learning methods to clinical conditions and was studying in that area in addition to conducting my primate experiments. I was also working as an intern in a clinical service off campus where I consulted on the application of behavior management techniques for developmentally disabled children. Unlike the colleagues who always had their eyes on job advertisements, I had never even looked to see what was available. I was not ready to leave, and I had not given it a serious thought until that day.

At dinner I mentioned the conversations to Kim, intending to express my lament that the unfortunate timing would make application impossible. However, her response was so dramatic that I didn't get to express the lament. I remember her staring at me with the quiet intensity of someone who had been told that her life-threatening affliction was now imminently curable, someone with a joy so deep that it bypasses the typical routes of facial expression. With the quality of a demand that rarely found its way into her voice, she said simply, "You are going to apply, aren't you?"

Kim had been separated from her mother Patty since her parents' divorce many years before, and Patty lived in Albuquerque with Kim's stepfather, Roger Skinner, who had recently visited us in Madison and encouraged Kim to communicate more actively with her mother.

The next week I met again with my mentors and expressed my interest in the New Mexico position. Harlow provided me with more information about the position. As it turned out, the position had opened up because David Bessemer—who had been one of Harlow's students—had decided to leave academia. He stated clearly that he wanted me to go there and follow through with what Bessemer had abandoned. He expressed a wish, not a demand. Nonetheless, with this motivation added to Davenport's confidence and support and Kim's wish to reunite with her mother, I knew I had to try to get the job at New Mexico. The three of us calculated that I could complete a dissertation-quality project, finish the required coursework, and take my comprehensive exams in time to graduate in the summer of 1971. Harlow and I had written a chapter together on the effects of early experience on primate learning that would be published soon, so I had the authorship needed to make my application to New Mexico credible. I formally applied for the position and was invited for an interview in late September.

To get ready for the interview, I talked to other graduate students who had been on interviews and, of course, spoke to Harlow and Davenport. Davenport was very familiar with Logan, as they both had been students of Kenneth Spence. He said that Logan was a very competent and intelligent person who had moved from Yale to New Mexico with the intention of creating a strong learning program, "a department to be reckoned with," as he had put it. So the question, he said, was whether I wanted to join a fledgling department or pursue an appointment at a more established institution. He said that he saw me as equipped to prosper in either situation.

Among other counsel, Harlow advised me not to drink to excess during the social parts of the interview and to make a point of volunteering to teach the introductory course, as most young faculty try to avoid this hard-to-fill assignment. He thought this would leave the important impression that I was a team player concerned about the responsibilities of the department and not just my own career development.

At the airport arrival gate I could easily identify Frank Logan based on the descriptions provided by Davenport and Harlow. He stood nearly six feet tall and had a deeply tanned, clean-shaven, handsome face. He wore a neatly pressed white guayabera and a western bolo tie with a brilliant turquoise stone in an intricate silver setting. He was the epitome of a southwestern gentleman scholar. Luckily I didn't appreciate the importance of these first moments of our encounter. In later years I heard him say that he could always tell whether he wanted to hire a prospective faculty member in the time it took for the person to walk from the gate to his waiting handshake.

The evening air was warm as we drove out of the airport in his blue Oldsmobile convertible with the top down. The Sandia Mountains glowed a watermelon pink to the east, and a deep-orange sunset was on display to the west. We drove to the Western Skies motel, which was close to Logan's large home in the fashionable Four Hills section of southeast Albuquerque. After I checked in, we sat in the lobby while he went over the events of the next two days. We made plans to meet the next morning and said goodnight. I felt very comfortable, with a heart full of excitement more than anxiety.

When Logan picked me up the next morning exactly on time, I was ready with a file of notes and data summaries and twenty newly minted slides of monkeys and graphs of my research data. On the way Logan explained that the department's classrooms and laboratories were temporarily scattered around the campus but would soon be united in a new building that would be completed by 1972. He parked his car and we entered Scholes Hall, a beautiful three-story building of classic adobe construction. Once in his office, he did not sit down or offer me a seat by his desk. Instead he invited me to examine a large set of blueprints lying on a solid brown rectangular table with a carved leaf border. He spent an hour showing me the building blueprints, pointing out where my laboratory and office would be located. This surprised me. I had not yet given my presentation and he was talking as if we already were committed to each other. It was becoming clear that the old-boys' network was in operation and this job was mine to lose.

The second stop of the tour was Logan's laboratory, located in a World War II–vintage wooden barracks a short walk from his office.

A sign identifying the building as B-1 hung below the army-green eaves by the entrance. Logan was clearly proud as we walked through the various rooms and he showed me his collections of rats, quail, capuchin monkeys, and rhesus monkeys. The facilities were primitive by any standard. The old wood rooms were dusty and unkempt, and the cages were dirty. Miniature stalactites of congealed urine, feces, and hair hung down under many of the rat cages. The drop pans under the monkey cages were filled with feces that overwhelmed the absorption ability of the sawdust with which they were filled. The rooms did not smell of animal life but of foul decay.

Logan seemed completely oblivious to these conditions. As we continued, he showed me the two large straight alley mazes that were the centerpieces of his lab. I thought of a remark Harlow had made about such apparatuses: "The straight alley maze begins nowhere and goes nowhere, but what happens in between is quantifiable." I decided not to share the observation with Logan, but smiled to myself instead. There was also an array of Skinner boxes and a data analysis room full of new calculators. As we walked through the lab, the graduate students working there gave me knowing glances, even as Logan failed both to acknowledge their presence and to introduce a single one to me. I concluded there was a degree of formality and interpersonal distance here that did not exist in Harry Harlow's lab.

Logan told me he wanted to show me where my temporary office and lab would be. On the way we walked into a long, narrow room occupied by nothing but a single monkey cage. Inside the cage, which was built for an adolescent less than half his size, was an adult male rhesus monkey, large by natural structure and incredibly obese. A bag of pendulous abdominal fat hung off his body like a large tan balloon full of water. As we approached the cage, the monkey began to spin upright, pointing his chin at the ceiling and flashing his large canine teeth at us while making cracking sounds in his throat. This odd movement was necessitated by the fact that he virtually wore the cage like a wire mesh suit. I recognized the open cuts on his upper arms as scars from self-mutilation. I knew this animal had been living in this condition for a long time. Offhandedly Logan said that I could have the monkey if I wanted him. The scene was horrific, but I felt that I was in no position to say anything.

The rest of the day went smoothly. The faculty members were friendly and positive about the department being born under Logan's leadership. Several said that it was a chance of a lifetime to help create a department from the ground up and that I would enjoy the opportunity. Later in the afternoon I gave my presentation to a filled seminar room. The slides and my accompanying narrative moved smoothly. I could feel my confidence growing as I answered questions, most of which I had predicted and specifically prepared for. The audience was attentive and applauded enthusiastically at the conclusion.

Logan then invited several faculty members to join us, and we went off to an Italian restaurant close to the campus. Dinner was filled with animated conversation, little of which was work-oriented or had to do with me or my presentation. It was a group that seemed to enjoy one another's company.

At one point, however, Logan, who was seated next to me, asked if I had any preferences about teaching. I had been waiting for this question. I responded that beyond courses in comparative psychology, animal models of psychopathology, clinical applications of learning, and primate behavior, I really wanted to teach the introductory course.

Logan's demeanor changed immediately. "What makes you think you know enough about psychology in general to teach the introductory course?" he said. "That is the course I teach. It will be many years before I trust you sufficiently to assign you to that course."

I was stunned and fumbled for a rejoinder. I was familiar enough with street power dynamics and primate social behavior to recognize that Logan's move was a dominance display. If I was going to take a position at New Mexico, I needed to know who was in charge. The situation called for signs of deference. I accommodated.

After dinner the group proceeded to Logan's home. Peder Johnson, a junior faculty member who had joined the department several years earlier and who was the chair of the search committee that had invited me, leaned close as we stood by the bar near the pool table and whispered that the real interview was now about to take place. I quickly came to understand what he meant.

Logan proceeded to form two-person teams to compete in an eight-ball tournament. If my performance under this contrived stressful situation was part of the interview, that was fine with me. Growing up in New York, I had spent many hours in pool rooms. In fact, some might say it was a significant part of a misspent youth. My assigned partner was Robert Grice, the department's distinguished professor. I quickly discovered that while he may have deserved his distinguished professorship, he was no pool player. Logan, on the other hand, was quite skilled and chose for his partner another skilled faculty member. Surprisingly, Logan also insisted we play for money. He directed each player to ante up five dollars per game. I had fifty dollars in my pocket, the full extent of my funds. I could not carry the scoring burden all by myself, and after an hour I was broke. When I reported the state of my finances to the group, Logan called an end to the competition and declared that he and his partner were the victors.

After borrowing twenty dollars from Peder, I challenged Logan to one game of straight pool to 125 points. While my demeanor was still appropriate and civil, I was angry and certain I could beat him at my favorite game. Logan declined. This part of the interview was apparently over.

The next day involved a very early breakfast meeting with other faculty members and an exit interview at lunch with Logan. During lunch at the Elks Club, I noticed that Logan and the others consumed a fair number of double martinis in short succession without showing any sign of effect. I thought of Harlow's well-known alcohol problems and wondered if this was a cost of academic prominence.

During the interview Logan told me, without room for negotiation, what the salary would be and to think about what equipment I would need for my laboratory in the new building. He made it clear again that I would be on my own as a primate researcher, with little help from other faculty. In fact, he envisioned that a part of my job was to stimulate interest in primate research among the faculty and graduate students.

The previous night's gaming ordeal forgotten, I readily identified with that mission. At the time I saw expanding the use of primate subjects in learning, physiological, and neuroscience research as one way to improve the quality and relevance of psychology in general. I saw myself encouraging graduate students and faculty not to shy away from the use of large-brained monkeys that belonged to the same taxonomic order as humans. I could see myself replicating Harlow's career on a smaller scale and creating a major national laboratory with dozens of primate researchers working under one roof to better the state of knowledge and the human condition. The opportunity that Logan was describing was the best that I could imagine. Logan was too good a psychologist not to see the energy and purpose that I was feeling as he described what my place was to be in his New Mexico endeavor. He knew he had me.

Some months later I read Franz Kafka's short story *In The Penal Colony*, in which a state official visits an island prison and is shown a machine that delivers punishment by cutting a statement of the prisoner's crime into his body. The visitor is horrified and disgusted but quickly concludes he is powerless to change things and agrees to his host's request not to openly speak against it. He later discovers that a new commandant who has been appointed to the prison has already ordered the procedure stopped. What I heard in the story was the all-too-frequent willingness of morally concerned people to stifle their points of view and undervalue the potential influence of their input. They just go along. I recalled the fat scarred monkey in the cramped cage, the mess in Logan's lab, the deplorable hygiene — and my silence about any of these issues. Surely I could have created a way to present my concerns, especially about the monkey, without overly offending Logan, but I didn't want to take that chance.

When I brought this up to some fellow students, all agreed that I was right not to speak out. Most landed, as I had, on a belief that the new facility under construction would take care of the problems.

* * *

As it was now clear that I would be starting my own primate labora-tory more or less from scratch, I knew there were many general skills I needed to acquire. I began paying a lot of attention to management issues, cage construction, and the medical interventions necessary to treat the common ailments and wounds that arise in a monkey colony. I spent time in the metal shop learning how to repair caging. In the electronics shop I learned the basics of constructing and repairing programming equipment. I also asked the attending veterinarian, Dan Houser, if I could go on rounds with him several days a week for my remaining tenure. He was surprised by my request but willing to have me along. He taught me how to anesthetize, clean, and suture uncomplicated wounds and how to select and administer antibiotics. I learned how to treat the common diarrheas as well as the more severe shigella dysenteries. He taught me how to clean teeth, and I assisted him when he surgically removed the large canine teeth from male rhesus to protect human handlers and the monkeys' cage-mates.

One afternoon I found Dan in the treatment area on the second floor of the lab. Lying on a brilliantly clean stainless-steel examination table was an adult female rhesus monkey. The reason for her presence in the treatment room was obvious: her abdomen was incredibly distended, the skin stretched taut like a drumhead. In my veterinary preparations I had read about gastric dilatation or "bloat syndrome" in monkeys, but I had never before seen an acute case. Her eyes were hazy, moving very slowly from side to side. Inside, gas was crushing her internal organs and interfering with cardiac function.

I helped Dan pass a nasogastric tube in an attempt to draw the gas and fermenting ingesta out of her stomach. It didn't help. He asked a staff member to call the investigator who was using the monkey and tell him to come to the room immediately. In the meantime he took a two-inch-long 18-gauge needle from his bag, cleaned it with an alcohol swab, and drove it straight into her stomach in an attempt to release the gas. It had no effect other than to create a smear of bright red blood on the shiny white skin of her abdomen.

The investigator, a graduate student, arrived shortly after being paged. Dr. Houser described the situation and the pain that the monkey was likely experiencing. The student asked whether there was any chance at all that the bloat would resolve on its own. Houser responded that the chances were close to zero and his preference was to humanely euthanize her. The student latched on to the slim possibility of resolution and rejected euthanasia. He explained that she was an important part of his research project and that losing her would create a serious delay. As I recall, the project was about identifying whether infants had a nipple preference during breastfeeding. The room remained silent as the procedures were repeated again and again until the light finally left the monkey's eyes and she died. The student left without a word, looking dejected.

The fact that something terribly wrong had happened hung in the room like stifling humidity. There on the polished silver table lay a female rhesus monkey that had been snared from her natural home in South Asia in the name of science. As we bagged the body up for disposal, Dan Houser said quietly to himself but loud enough for me to hear, "The animals serve us, and we must in turn serve them at times like this."

This was the first time that I heard someone in authority describe what amounted to an ethical principle regarding researcher responsibilities to laboratory animals that went beyond the dodge that animal use was justified if humans could not be ethically used. Dr. Houser was stating that there were limits to what we could legitimately extract from animals in research. There was a point where the advancement of science had to give way to the needs of the animals, and that point had been ignored in the current case.

Why was this a surprise? After all, I was familiar with the decision my family made when our dog Penny was suffering from distemper, which rendered her sick and trembling. It didn't matter how much we wanted her to live and stay with us; the issue was her suffering, and we asked the vet to end her life for her. I knew the monkey in question was valuable, not for bringing pleasure to the lives of lab personnel but for what she might be able to reveal to us about primate behavior and, by inference, perhaps humans as well. I struggled with why that difference trumped basic decency and respect. I was soon faced with a similar issue in my own research.

My testing involved evaluating how isolate and socially experienced monkeys performed on so-called schedules of reinforcement. To ensure that the monkeys would be interested in performing the assigned tasks, they were deprived of food or water or both for twenty-three hours before the test. We either shut off the water valve that provided water to them or put a sign on the cage asking the staff not to provide the normal food. This signaled that feeding or watering was now my responsibility.

One afternoon my name was called over the paging system—I was to report to the treatment area in the basement of the lab. When I arrived, Dr. Houser had a stethoscope to the chest of a small monkey. I assumed I had been paged to observe an interesting clinical case, as I was no longer testing any animals in that part of the lab. When I got up next to him, however, I could see the monkey was not young but emaciated into nothing but skin and bones.

I immediately knew what must have happened. After this female monkey had finished our response tests, the sign prohibiting the staff from feeding her had not been removed. As a result, the animal had not been fed for weeks and was close to death.

In my fear and embarrassment, I blurted out that Scott, my undergraduate research assistant, had screwed up. Houser turned his head away from the monkey and looked at me for a few seconds. There was no emotion on his face, and his lips were pressed together in a straight line. Without a word he returned to his clinical efforts.

His stern look told the truth. I was responsible. The monkey, whom I had not named, stopped breathing several times as we worked to revive and stabilize her. Eventually Houser's efforts were successful, however, and she was placed in a holding area where she could be observed and tended to twenty-four hours a day.

For more than a month I visited her several times a day to do what I could, and she eventually recovered to the point of looking somewhat like her normal self. Surprisingly, no one officially chastised or punished me, perhaps because my remorse was so visible. When she was returned to the general colony, I added a piece of tape to her cage and printed on it the name "Miracle."

* * *

During my final year at Wisconsin, I had one more opportunity to take part in one of Harlow's research seminars. During this seminar on the history of psychology, Harlow revealed several aspects of himself that are relevant to understanding how he thought about his work. The seminar was organized around presentations given by graduate students about notable figures in the history of animal behavior studies and psychology. We covered the ethologist Konrad Lorenz, the early psychology figures Edward Titchener and William James, and more contemporary individuals in learning like B. F. Skinner and Kenneth Spence. The presentations were organized loosely so that Harlow could fill in with comments. This format suited Harlow perfectly: he liked to sit seemingly uninvolved and on the edge of discussions while he was in reality quietly poised to interject hard, direction-changing questions or humorous rejoinders. In fact, Harlow did a lot of his teaching this way, making brisk comments that left you thinking about their meaning long after their delivery. Mostly they were offered as intellectual gifts that could be accepted, questioned, or ignored. Speaking for myself, I often wished that he would follow up his comments with explanations, but he rarely did. I was generally reluctant to ask him questions about his points in fear of revealing that I was not smart enough to get the meaning the first time around.

During the discussion of the work of Konrad Lorenz, there was puzzled amazement among the students when the presenter, Charley Weisbard, indicated that Lorenz believed that a researcher needed in a sense to "fall in love" with the animals if the research was to reveal an in-depth understanding of the animals' behavior. In this context Weisbard also emphasized the concept of *umwelt*, a term, derived from the work of the biologist Jakob Von Uexkull, that refers to the perceptual world of the animal—and the necessity of the researcher's entering it. We wondered how such an attitude and approach could avoid what we considered the deadly specter of anthropomorphism and anecdotalism that in our minds meant nonobjective science.

Harlow was uncritical of the implications of the Lorenzian method. He did not reject it as I expected he would. I took this as a demonstration of respect for a scientist with goals different from his own. His work was not directed toward learning about monkeys as monkeys; he wanted to use them to model human phenomena.

However, Harlow did not invite us to go further in thinking about what it might mean to try to enter the perceptual world of our experimental animals. Nor could any of us imagine choosing to do so. Our collective refusal to pass through this threshold illustrates the ethically sterile and purely utilitarian attitude toward animals that prevailed at the time. It is worth considering whether confronting the actual internal experience of a baby monkey being forcefully detached from its mother's embrace and forced to live in a constantly illuminated stainless-steel box for months would have caused any of us to think in terms of ethical limits of harm. Speaking for myself, I think not. It is more likely that such an exercise would have resulted in more ideas about other experimental models of human functioning and suffering.

Throughout the seminar Harlow gave examples of what he saw as genius and told stories illustrating the foibles and blind spots that plague all researchers at some point. In the realm of genius, he extolled the unique creativity of William James, describing him as "the kind of man who would look like a stranger in any country." As for blind spots, he used as one example Kenneth Spence, heir to the Clark Hull learning empire, who had died four years earlier. Harlow saw Spence as fundamentally brilliant yet possessing flaws that Harlow clearly wanted us to avoid. He pointed out that Spence impaired his own chances to become a more important player in psychology by acting in a condescending manner and frequently displaying a bad temper. He put it this way: "Spence saw his critics as being composed of two types, either misguided or stupid." As for his temper, Harlow told us that Spence "was not mean except when he was drunk; however, he was drunk from 1951 to 1964." He said Spence dogmatically believed that animals do not possess the kind of mentation that is capable of insight learning, and that all learning is incremental and occurs when a required "correct" response is moved into prominence once an organism comes to expect reinforcement. Harlow strongly disagreed; he was convinced that monkeys show strong evidence of complex cognitive and emotional capabilities and are completely capable of insightful behavior. Harlow admitted that he had devoted a good deal of his career to "destroying" Spence, a path that he abandoned after he married one of Spence's students, Margaret Kuene, and she told him he was "a damn fool" to spend his time that way. I took this comment to mean that he, too, had a blind spot: he had fallen into the trap of competing for dominance rather than just trying to understand the mechanisms of learning.

According to Harlow, Spence made a "fatal blunder" during an American Psychological Association (APA) meeting that Harlow also attended. A presenter challenged Spence to explain how a set of findings from human subjects that he had just described could be explained by Spence's theory. Spence got to his feet and claimed loudly that he didn't care about the human data because his theory was based on rats. For Harlow, building a theory to explain nothing but the behavior of rats was a major mistake. The study of animal behavior makes sense, according to Harlow, only if it helps to explain human psychological functioning. In that context, he went on to predict that Spence's theories would "disappear." This prediction was certainly accurate. The Hull-Spence approach now lives on only as a chapter in courses in the history of psychology.

One of the revelations Harlow made during the seminar went a long way toward explaining rumors I had heard about him since the beginning of my time at Wisconsin. It was intimated that Harlow battled with bouts of severe depression and that this was the cause of both abusive drinking and frequent periods of absence from the lab. I had heard that one particularly long and deep period of depression had occurred on the heels of Harlow's receipt of the National Medal of Science at a ceremony at the White House in February 1968. When I heard about it, this had made no sense; I couldn't understand how receiving a prestigious award could bring on depression. Then, during the seminar, Harlow explained that over the years he had found that receiving an award or honor was a double-edged sword. One edge helped to slice away the insecurity that his work was in reality unseen and unappreciated. The other edge assaulted his confidence by suggesting that perhaps his creative days in science were over. The National Medal was particularly threatening in this regard because it was usually given at the end of one's scientific career, signaling, in Harlow's mind, the impending scientific death of the recipient. For him, being scientifically dead was the same as being biologically dead. He indicated that he had managed to emerge from the depression prompted by receipt of the National Medal only after committing himself to proving that his creative career was not over. As it turned out, this was the beginning of the career phase in which he attempted in earnest to model specific psychiatric disorders, particularly depression, by exposing monkeys to harsh living conditions.

At the end of one class period, I approached Harlow individually as he sat quietly smoking the last of a cigarette. I observed that he hadn't ranked his own scientific longevity in comparison to Skinner and the others, and I was interested in hearing where he would put himself. Before responding he seemed to check to see if others were listening. He then looked at me squarely and said that his approach to research was, or had become, something akin to being a "stripteaser." I asked him what he meant. He didn't answer but instead smiled, stood up, and left the room.

I have kept his comment to myself and thought about it often through the years. In that moment in 1971 I sensed that his guard was down and his professorial showmanship was not on display. Harlow rarely used words without specific intention. Perhaps he was simply saying that he tried to capture the attention of other scientists by making brash and controversial proposals. But a striptease is not really about creating a surprise. All observers of a striptease artist's show know in general where the performance is headed; the question is just how provocative it can be and how well the performer can increase arousal by encouraging self-involved fantasy. In other words, the observer is pulled in and transformed into a participant.

I have come to believe that Harlow's "striptease" metaphor is best understood in the context of his advice to "think photographically." Images, unlike data, pull the observer into the predicaments of the monkeys in the photographs. The observer readily imagines the emotional or behavioral state taking place and how it was produced. This is why the mother-surrogate experiments were so successful and why they continue to this day to be covered in all introductory psychology textbooks. The images of monkeys avoiding a cold, food-providing wire "mother" and holding on tightly to a warm and soft terry-cloth-covered one communicate important elements of the nature of attachment virtually without the necessity of words or numeric data. Like a striptease act, they kindle the observer's imagination and in that way make their impact. Ultimately, however, this strategy carries a risk. If photographs of research results depict suffering, observers can identify too strongly with the suffering and be jolted out of the desired mode of involvement, just as observers of the stripteaser's show can move, to the artist's distress, beyond the realm of fantasy. Harlow's late-career ethical downfall occurred because he pushed too hard against the risk inherent in his approach.

Harry Harlow was in important mentor for me because he consciously strove to shape his students as thinkers and intellectuals, not just as researchers in psychology. Like no other teacher I had at Wisconsin, he shared a knowledge of literature and art and, when discussing the work of scientists, frequently mentioned the artistry of such performers as the Polish pianist Ignacy Paderewski and the prima ballerina Anna Pavlova. The library in his large Primate Lab office was full of the great books and examples of his own pen-and-ink drawings and poetry, not just psychology texts and technical papers. He clearly modeled for us that an academic psychologist needed to be a person of letters and not just science. He reminded me that the "Ph" in the PhD degree was there for a reason. It was an indication that the holder was a *philos sophia*, a lover of wisdom, not just an expert in psychological research. This point made a great impression on me and played a part in my later decision to study the discipline of moral philosophy and change my research life.

* * *

By August of 1971, I had completed all my requirements, including the defense of my doctoral dissertation, "An Operant Analysis of the Effects of Differential Rearing in Rhesus Monkeys." What remained to be done before I left for New Mexico was to negotiate with Harlow which monkeys he would allow me to take with me so that I could continue my research.

When I met with Harlow, I brought a list of identifying numbers of the animals that I had used in my dissertation plus those of two others, a young pair of stump-tailed macaques. I had befriended these monkeys and then rescued them from participation in a brain-lesion study by claiming their involvement in my own work. Harlow said that it was fine with him if I took these monkeys but that he would check further to be certain. He said I needed to ask Ken Schiltz for "the key."

I did not know what he meant but did not want to bother him further. So I went to Ken's office and told him what Harlow had said. He reached into his desk and retrieved a key. I followed him to a large storeroom some distance from the lab that I had not noticed before. Ken opened the door and invited me to enter. Inside was a large collection of retired caging, testing equipment, transport cages, catching nets, and examination tables, all very serviceable. Ken told me to select what I needed to get a start at establishing a small lab at New Mexico.

I recalled that during one of my seminars with Harlow he had made the comment that the first year as an assistant professor was harder than the entirety of getting a PhD. He spoke of having no facilities when he had arrived at Wisconsin in 1930, as the lab he had planned to use had just burned down.

With these difficulties in mind, he had created a way for his students to get off to better starts. I felt privileged to receive this gift and participate in this Harlow student tradition.

My last day in Madison involved packing, participation in a lab softball game, and attending a send-off party at Elaine Moran's home. It was a classic lab party with laughs, beer, bratwurst, and cannibal sandwiches — the latter a Sheboygan specialty. I loved these people, and the atmosphere that evening made it clear that they cared for me as well. The gift I received, presented to me by Steve Suomi, was two softballs autographed by fellow grad students, faculty, and staff. Characteristic of the lab social structure, Harry Harlow's and John Davenport's signatures had no prominent place but were intermingled among the rest. Suomi joked that given my destination, I was going to need as many balls as I could muster.

Before going home, Kim and I made one more trip to the lab. Using my key, I went to the treatment room on the second floor. I opened the drug cabinet and took one bottle of ketamine hydrochloride, a dissociative anesthetic used to restrain monkeys so minor medical procedures could be accomplished without using full anesthesia; a bottle of pentobarbital, a concentrated barbiturate used to induce deep anesthesia and in large doses to euthanize animals; and a variety of different-size syringes. I left a note on the shelf acknowledging the withdrawal. I knew that this was a questionable move and hoped that it wouldn't put anyone in a difficult situation, but I had my reasons.

I then went to see all my monkeys for one last visit. I turned on the lights in the holding rooms where they were housed. As I walked through, all the monkeys were either asleep or quietly awake. I enjoyed seeing them all peaceful and huddled into themselves and not responding to a human's presence with fear grimaces, threats, door-slamming, and screeches. I made a special point to visit the two stump-tailed monkeys, Manny and Greta, that I wanted to come with me to New Mexico. They were really my pets, and I feared that another researcher would snag them before I was able to get them there. They barely opened their eyes as I stood in front of their cage, and they did not release their mutual sleepy embrace. I whispered a promise to them that I would get them as soon as possible.

As I left the second floor, I waved goodbye to Susan DeLuna, who was working the night shift in the nursery. I put the drugs and syringes into the Styrofoam ice chest in the trunk of my car. The next morning the movers picked up our personal belongings and the lab equipment that Harlow and Davenport had released to me, and we left for Albuquerque.

3
Practice

No doubt, no awakening.

C. C. CHANG

After two days of driving, Kim and I approached Albuquerque from the east. It was late afternoon when we drove through Tijeras Canyon, flanked on both sides by the steep face of the Sandia and Manzano Mountains. The sun was in our eyes, and the air coming through the open car windows was very hot as we crested the rise on the edge of the city. At that point the closeness of the canyon walls gave way to the Rio Grande Valley, a vast expanse of aridity that went on for miles in all directions. I pointed out the Western Skies Hotel, where I had stayed the year before. Kim and I looked at one another and said spontaneously in almost choral unison, "Here we go!"

We drove directly to Kim's mother's house, where we were received with excitement and real warmth. Kim and her mother fell into each other's arms and cried. This made me feel that we were in exactly the place we needed to be, both personally and professionally. After a meal, we drove to the apartment that Kim's mother and stepfather had rented for us. It was roomy and clean, a short distance from their house, and freshly painted white and turquoise. We unloaded our belongings and went to bed, hardly noticing the stifling heat.

The next morning we divided the work just as we had upon our arrival in Madison; Kim was left to arrange the apartment while I went to the Psychology Department office.

As I entered, Eleanor Orth, Frank Logan's secretary, looked up and without hesitation said, "Oh, Dr. Gluck, we have been waiting for you; we are so glad that you are here." She was the first person outside of my Wisconsin friends to call me "doctor." She quickly had me fill out my book order forms for the two classes I was teaching and gave me a set of keys to my office and lab space.

After completing the paperwork I went to my office. B-1, the temporary building where it was located, seemed even shabbier than I had remembered. My office contained an old brown wooden desk, a creaky and unstable swivel chair, a long table, two army-green file cabinets, a heavy black phone, and a wooden coat rack. The room looked as though it had been in the center of a dust storm and had a musty, almost acrid smell.

I sat in the chair and examined the space more carefully. There was a large green blackboard on one wall with a series of algebraic equations written in yellow chalk; I recognized them as statistical equations for calculating variance and the standard error of the mean. A large dirt-streaked window was in the east-facing wall, providing a view of the parking lot and a small, beautiful adobe classroom building surrounded by trees and vines. I looked through the desk drawers and file cabinets and found them totally empty. Frankly, I was hoping for a note or letter of some kind from Dave Bessemer, the faculty member I was replacing. However, as I got up to leave, I spotted a small carved and delicately painted wooden monkey hanging by its up-stretched arm and curved hand on the coat-rack hook. It was the only object in the office that had been recently cleaned of dust. Perhaps this was a gift from Bessemer, a wish to me for good luck? I chose to think so.

After leaving the office I went to see if the rhesus monkey that had been in the adjoining room when I interviewed months earlier was still there. He was, and nothing had changed. His cage was still a wire-mesh trap, and the cuts on his arms from self-mutilation looked worse. I examined the monkey's arms more closely and could see the scar tissue and open wounds arrayed like pink-colored scrambled eggs just below his shoulders to his elbows. I offered him a piece of the uneaten sandwich I had with me, and he grabbed for my arm instead. He shook his cage violently while continuing to try to get hold of my hand.

Leaving, I drove back to the apartment and retrieved the drug bottles and syringes I had taken from the Primate Lab the night before we left. Kim was actively involved with arranging furniture and hanging pictures. I hardly exchanged a word with her as I gathered the material and left. I returned to B-1 and asked Bob Paul, one of Logan's graduate students with whom I had become friendly during my interview, to help me. We went into the monkey room. I asked Bob to draw the monkey's attention with some treats while I circled around behind him. His meaty thigh, pressed against the wire mesh, was my target. I took the 1-cc syringe that I had filled with ketamine and jabbed him with it, quickly emptying the contents before he could turn and defend himself.

Bob had no idea what I was doing. Within just a few minutes, the large, soft monkey began to slowly lean to one side of the cage. In a few more minutes his body was motionless while his eyes danced. I poked at his shoulders through the mesh, and he did not respond. He was ready.

I opened the front door to the cage, removing the hasp and clipping it to the side of the cage. With bare hands, I reached in, took hold of his upper arms, and pulled him carefully backward through the door. It took some maneuvering to get his body through the opening. Because of his heft I carried him with some difficulty to a wooden table and laid him on his back. His eyes bounced irregularly from side to side as I scanned and poked his body, looking for an accessible vein. The only easily available one that I could find was the great saphenous vein, running up the backside of his right calf muscle. This was not ideal, but it was going to have to do. I retrieved the 10-cc syringe full of pentobarbital and inserted it lengthwise into the vein, being careful not to penetrate through the other side. I slowly emptied the contents and withdrew the needle, stopping the flowback of blood and drug with my thumb.

After just a few seconds the monkey's eyelids began to close, and the depth of his breathing became more and more shallow. Soon there was no sign of breathing. Having forgotten my stethoscope, I put my ear to his chest to see if I could detect a heartbeat. The fine hair on his chest tickled my cheek, and the soft warmth of his body felt comforting. I could hear no heartbeat. I pinched the soft tissue between his toes hard, and there was no response. I touched his eye and there was no blink reflex. He was dead.

Bob, who had left the room, returned with a plastic garbage bag. We placed the enormous monkey into the bag and tied it closed. Bob said that he would take care of the disposal. I thanked him. He responded with something like "No, no, thank *you*. Couldn't stand to look at that monkey anymore." As he turned to leave, he looked back at me and asked if Logan knew about this. I said no. "He might not ever notice," he said.

I went back to my office and thought about what I had done. Did I kill this monkey to make living in B-1 more tolerable for me? With him gone, I no longer had to be reminded of his horrible condition and my reluctance to face Logan directly about his failures as a lab director. Was it just a behind-the-back shot at Logan's authority? Or did I do it for the monkey, whose life had been so twisted by his treatment that he had no prospects either as a research subject participating in an experiment or as a being living out his complex intellectual and social capabilities? I honestly could not disentangle these motives. While they all played a part, I just thought that there was something right about euthanizing the monkey. Some amount of pain and anguish in the world had been eliminated—the monkey's and mine.

Twenty-eight years after this incident, I was working with colleagues in the medical school on a project when the staff statistician and I were left alone for a few minutes waiting for the others to return. I recognized him as having been the undergraduate lab assistant at the time I first arrived at the department. We began to fill one another in on the whereabouts of the students and lab technicians from that time. I asked him if he ever thought about the work he did for Logan back in the day.

He responded immediately, saying that he knew most of the

animals he worked with were "just rats" but was sure they "must have had some ability to suffer." His recollection then turned to the monkey in B-1. He said he still thought of him in his small cage and lonely environment. Grimacing, he recalled that the monkey had been removed from his confinement shortly after I came to the lab. He looked at me eye to eye, hesitated, and said, "Thanks for helping that guy out." I nodded.

Over the next week I moved my books and papers into the office and began to feel at home. I introduced myself to more graduate students, had several lunches with Peder Johnson, and worked on preparing for my classes. I also began writing, with Douglas Ferraro, a grant in the area of drugs and behavior. Ferraro and I had maintained a correspondence since my interview, and we had decided to establish a research collaboration once I arrived. He had an established lab where he studied the effects of drugs of abuse on learning. While most of his research utilized rats and pigeons, he also worked with a group of chimpanzees that were kept on the Holloman Air Force base in Alamogordo, New Mexico. Many of these animals had participated in Air Force / NASA–sponsored research on putting a human being in space. Once the Air Force was finished with them, they leased them out for use in biomedical and behavioral research.

Although I had done just a little work in psychopharmacology at Wisconsin, and it was not of great interest to me, I reasoned that it would be a good addition to my basic understanding of the effects of early experience on learning. I found Ferraro to be a very capable and ambitious psychologist with a Skinnerian academic pedigree from Columbia University. I enjoyed his company, and I felt that we would work well together.

One afternoon, about two weeks into the semester, I received a call from Philip Day, the campus veterinarian. He asked if I planned to bring any monkeys onto the campus before the new building was complete. I said that I expected to have some monkeys before the end of the semester. In that case, he said, we should meet to discuss housing options. He suggested that we meet in his office on the Medical School campus. He proved to be a pleasant person with a clean-cut military bearing. His hair was blond and short, his body fit, and his interactional style friendly but no-nonsense. When I arrived, the first order of business was a tour of his animal facilities. What I saw was distinctly different from the facilities on my side of campus.

The rooms had glassy epoxy-sealed floors, walls, and ceilings and stainless-steel equipment. The surgery was built to human standards and was brilliantly clean. Similarly, the animal rooms and cages were spotless. In one room mixed- and full-breed dogs whined loudly while wagging their tails at me from individual cages stacked two high. I asked Phil where they came from, and he said the City of Albuquerque dog pound. In other rooms, rats and mice sniffed the air through clean wire-mesh cages. There were few odors that would identify the place as an animal facility. Technicians in clean and neat monogrammed blue uniforms moved purposely between animal holding and procedure rooms. I was envious but could imagine that when the new psychology building was complete, my facilities would be of equal quality.

After the tour we sat in his well-organized windowless office. His first emphatic points were that there was no space in his facility for any monkeys that I might bring to campus and that he did not have any extra caging to loan to me. Having made that clear, he went on to say that he had a great deal of primate experience as a vet in the Air Force and had worked with primates for the space program at Holloman Air Force Base in southern New Mexico. Finally, he said the university was in the process of forming an animal care committee that would inspect all the animal facilities on campus and make a report to the administration twice a year and asked me to participate as a member.

I left the meeting encouraged that there was an experienced primate vet on the campus. Phil's presence would make it more feasible to get my lab started as soon as I could find an appropriate place to house the monkeys from Wisconsin. When I talked to Frank Logan about the invitation to serve on the animal care committee, he didn't just clear my participation but insisted on it.

Shortly after my meeting with Phil Day, I made an appointment to talk to Logan about space for a small primate lab. In earlier discussions he had made it entirely clear that it would be a mistake for me to wait until the new building was complete before getting my research program off the ground. Now I told Logan that I would not bring the monkeys that were waiting to be shipped to me from Wisconsin to my assigned lab in B-1. I was taking a chance in making the request to house the monkeys elsewhere, since it surely demeaned his decision to keep his own lab in B-1. He made no argument, however, and said that he would get back to me about other possibilities.

Within what seems now to have been just a week, I received a call from Guido Daub, the chair of the chemistry department, who offered me space in his building to house a small number of monkeys. We met to work out the details, and he took me to the basement to show me the location he had in mind. When we got there, we walked into a spacious unoccupied laboratory that was fitted with standard chemistry benches and fume hoods. Exhilarated, I immediately began to imagine where I could put cages and the testing equipment. But Daub continued to walk through that lab, through another door, to an adjoining barren thirty-foot hallway. When I caught up with him, he pointed to a room at the end of the hall and said, "Take a look." I walked down the hall and opened the heavy metal door. Professor Daub flipped on the light switch. The room was a dimly lit concrete box fifteen feet square with no windows, no access to water, no air vents, and no floor drains. One small light fixture was centered in the ceiling. It looked like it could have been constructed using the plans for World War II German Siegfried line fortifications. My impression was supported by a small sign attached to the light switch that read "Explosion Proof."

Professor Daub must have read the disappointment in my face, as he began to list the virtues of the space. He pointed out that it was off on its own and away from noise and human traffic, that I could use the hall for my test equipment, and that there were lockable doors on both ends of the hall. I agreed and said I was very appreciative of his efforts, but I had hoped to have a little more room and a few simple amenities like air circulation and running water.

Daub then asked me to sit down and proceeded to tell me the story of the discovery of electromagnetic waves by the physicist Heinrich Hertz. I was confused by this apparent tangent, but he was a good storyteller and quickly got to the relevant point. He explained that Hertz's incredible discovery was basically due to sheer luck because the small space he had been assigned for the work by his department chairman turned out to be just the right size and configuration to make detection of the waves possible. He suggested that perhaps I would be just as lucky.

His humor and wisdom won me over. I now had a lab that was far better than what I'd had at B-1, and I would find a way to work with its deficits. I knew I could work out procedures that would make it possible for the monkeys to survive in the space, and most importantly, I could begin to collect data. Professor Daub gave me a set of keys and wished me luck.

The next week I was lecturing about the importance of research and the creation of new knowledge to my new class in comparative psychology. In this context I described my plans to establish a primate lab and the work I intended to pursue there. At the end of class one of the students approached me and said that he would like to help me build my lab. He explained that he didn't know much about animal behavior but was fascinated by the topic. He was working at the Lovelace Research Center in southeast Albuquerque and was associated with a primate-learning project there, but he was not working with the monkeys directly. He let me know that he was older than the average student, had been recently discharged from the Navy, and wanted to get involved "hands on" as soon as possible.

I recognized my own student enthusiasm in his face and could see clearly his guileless honesty. I remembered the meeting with Dr. Davidson that had kicked off my own research career when I was an undergraduate. On the spot, Jeff Sproul became my first research assistant.

Jeff was a man of his word. He said he was not afraid to get his hands dirty, and he proved that he wasn't. We used his father's 1960 red-and-cream-colored Ford truck to move test equipment and the cages stored in B-1 to the chemistry building and to pick up all manner of supplies. It felt like I was reliving my first few months as a graduate student in Wisconsin as we fabricated a lab out of spare parts and Harlow's startup gifts. Jeff and I worked long hours getting everything ready for the first shipment of nine monkeys that was about to arrive from Wisconsin. Together we located sources of wood shavings to use in the drop pans under the cages, discovered a ranch feed store from which we could buy monkey chow, secured surplus electrical-programming equipment from the university's salvage department, and purchased medications from a local small business named the Great Western Serum Co. The managers at the feed and medical stores seemed interested in hearing about my plans and were enthusiastic to help a new UNM professor in any way they could. They lowered prices, provided large amounts of free samples, and at times "forgot" to charge me at all.

We set up my test equipment, created a medical treatment area, and scrounged a refrigerator at a garage sale to facilitate food preparation. When the lab was complete, I invited Dr. Day to come and inspect my nascent operation. He was surprised at what we had been able to accomplish and offered to be on call for us when veterinary medical needs arose.

Just before the monkeys were to arrive, I drove out to the feed supply store to pick up the commercial monkey chow I had ordered. As I was loading several bags into the back of Jeff's truck, I noticed what looked like a monkey living in a wood-and-wire enclosure in the corner of the property about fifty yards away from the store. Walking closer, I saw that it was a large dark-brown male stump-tailed monkey. I was shocked. He was older than the two stumps I had left back in Wisconsin. He was sitting on a perch about eight feet off the ground outside a flimsy house constructed of plywood with a few asphalt shingles arranged irregularly on the roof. Several twisted and dirty ragged blankets protruded from the door of the house. I made some lip-smacking noises to get his attention. He looked down on me, then swung off the perch, grabbing the chicken-wire wall and climbing down to my level. I squatted down close, and he lip-smacked back to me. He pressed his chest to the mesh, lifting his face skyward in a clear invitation for me to groom him. I was careful not to put my hand completely through the mesh in case he was luring me in with intentions to grab me. I scratched his chest for a few minutes, and he never moved. When I stopped and withdrew my hand, he stared at me and made an aggressive throaty growl, as though he was saying that he had not given me permission to stop.

I returned to the store and engaged the owner, a thirtyish man with short blond hair dressed in full cowboy attire, in conversation about the monkey. He told me that he had purchased the monkey at an animal auction in west Texas about three years earlier. He did not seem to know much about his history and was unaware that he was a rare and endangered monkey from southeast Asia. I asked if he was for sale, and he said no with a very emphatic tone.

Each time I went to pick up monkey chow thereafter, I visited this monkey and began bringing him fresh fruit and peanuts to eat. Soon he recognized Jeff's truck and would be on the ground waiting for us. We went through the grooming ritual even on winter days when the temperature was freezing. In the back of my mind I thought that once I was in the new building I might make a serious attempt to buy him.

The Wisconsin monkeys arrived on an American Airlines flight early on a Saturday afternoon. Jeff and I were there waiting for them. When the stainless-steel transport cages were delivered to the air freight section, I ran over to the loading dock, anxious to see if they all had made it safely. I had expected nine monkeys, but when I counted the transport cages there were eleven. While Jeff dealt with the paperwork, I checked each cage, looking through its barred window, and was relieved that all the monkeys seemed frightened but alert. I recognized each one as I went from cage to cage. The unexpected monkeys were the two stump-tails, Manny and Greta. To me they all looked like old friends, but I was particularly happy to see the stumps. I guessed that they were sent ahead of time because there was a threat of their being used in an invasive experiment. When I called the Primate Lab later that day, I found that this was indeed the case.

I delivered orange sections to each monkey and was further comforted when they ate them immediately. When Jeff joined me, I introduced each monkey to him. He had previously listened to my detailed descriptions of them and tried to guess who was who. We loaded them all in his truck and drove to the lab. Before going to the chemistry building, Jeff and I went to B-1 and retrieved a large double cage that I had not expected to have to use to house the two surprise stumps. The chemistry holding room would now be very crowded. Using transport cages, we removed the monkeys from their travel cages and released them into what would now be their home cages.

Once in place, the isolates immediately began to display their stereotyped behaviors. E-15 spun clockwise while standing bipedally, E-7 did repeated backflips, and E-12 bit his arms. The socially reared animals were curious but apprehensive, holding on to the sides of their cages while visually exploring the surroundings. I waited to see that each monkey ate something and drank some water before I locked the doors and returned home.

Kim and I toasted their arrival with some champagne that we had been carrying around since our wedding day. Before going to bed that night, I went back to check the monkeys, refill their water bottles, and turn off that single light bulb in the ceiling. When I opened the door, I could see that they all had settled down and were seated on the small narrow perches eight inches off the floor of each cage. They looked great, and I was proud of myself.

I knew that Harry Harlow would also be proud, and that meant a great deal. I felt that I was reliving my version of the story that he always told about his own beginnings as an animal researcher: arriving on campus and finding that there were no animal facilities; chiseling with his own hands the concrete pillars that obstructed the usable floor space in the small building he took over, which was not designed to be an animal facility; and funding his initial research initiatives out of his own pocket. While I never thought for a moment that I would ever attain the scientific stature of my mentors, I knew that I had much to contribute to understanding the effects of early experience on learning and primate behavior more generally, and following Harlow's path added to my certainty about that.

Nevertheless, the facilities and housing arrangements in the basement of the chemistry building were crude and barely adequate. I would describe them with the wisdom of hindsight many years later when I wrote the following in the foreword to the 1997 edition of

Comfortable Quarters for Laboratory Animals, published by the Animal Welfare Institute:

> The comfort of the cages for the animals was not a definitive concern. The reasoning was rather simple: the monkeys were more physically flexible than was the metal and concrete space that was available to me. In other words, the structure of the rooms dictated the animal treatment. This utility-oriented perspective did not reflect, I think, a disinterest in the animals. Rather, it represented a kind of blindness brought about by my unitary and simplistic focus on the concept of experimental control and a reluctance to attribute feeling states to the animals. From this perspective what mattered in animal housing was whether the cages were uniform and "standard." Whether the experimental animals were comfortable in them was not the main point. As long as the animals were not obviously harmed by the housing (i.e., they ate, stayed hydrated, passed firm stools) in the short term, and the cleaning and feeding functions could take place efficiently, the circumstances were judged to be adequate.

If there were precursors of ethical concern present at the time, they were hidden by the romance of the moment and my state of self-congratulation. While I could readily see the ghastly predicament of Logan's monkey, evidence of the harm that my animals were already suffering was invisible. The stereotyped behavior, the occasional self-injury, and the difficulties in adjusting to novel environments were all parts of the "isolate syndrome" that was created purposely for research purposes. The fact that these behaviors originated from an experimental plan placed them in a category different from the suffering that I saw in Logan's monkey, which was the result of gross neglect.

Jeff and I split the duties of opening the lab in the morning and shutting it down at night. For some reason Professor Daub would not let me dismantle the light switch and install a timer to at least get the lights on and off without our having to be there. I think he believed that if I modified the structure in any way, it gave me a more long-term claim on the space. That was the last thing he wanted to deal with, as he was starting to get complaints about foul odors from the faculty on the upper floor. Twice a week Jeff and I met very early in the morning to clean the drop pans. We carried them out to behind the building, where we dumped the waste into plastic bags. We then washed the drop pans, using a hose that was conveniently located on the building wall. It was not possible to wash down the cages in the room, because there was no floor drain. Instead we removed the monkeys, briefly holding them in transfer cages, and scraped the dried feces from the cage floors. We then scrubbed them with disinfectant, rinsed them, and let them dry before we put the monkeys back in them. We continued this schedule until the psychology building opened one year later.

After a few weeks I was able to continue my experiments examining the impact of standard food rewards on the behavior of socially deprived and control animals. Jeff quickly joined that part of the work, and collectively we were getting some interesting data that confirmed earlier evidence that socially deprived monkeys require more frequent rewards to maintain their attention to learning tasks. It appeared that something had been damaged in either their neural reward system or their attentional system. Of course I took copies of the data sheets and graphs to Frank Logan so he could see that my lab was up and running.

As the end of the spring semester approached, I was beginning to worry how I would be able to financially negotiate a summer during which I was not on contract. I was disappointed to learn that there were no summer teaching assignments available, and the grant application that Doug Ferraro and I had submitted had not yet been reviewed. At the point of my highest concern and lowest mood, I received a letter from Harlow offering me a four-week summer appointment back at the Wisconsin Primate Lab to help him with a number of unfinished manuscripts. The honorarium he offered was roughly four times my monthly New Mexico salary. I knew that this offer was not really about helping him but rather an example of his unofficial welfare program for recently graduated PhDs. He knew I could use the money and some time among old friends and colleagues. It was his way of expressing the message that once gone from the Primate Lab: you were not forgotten and were still an important member of the Wisconsin lab community.

Jeff agreed to cover the monkeys for the month of June, and Kim thought it would be good for me to go. I accepted the offer and did so again the following year.

* * *

Part of what it meant for me to become established professionally at the University of New Mexico was to have a laboratory and support staff that did not depend on the generosity of others. When the psychology building opened in August 1972, I was provided a research home of my own that had been constructed with the conduct of animal research as its primary focus. After dismantling and moving my lab to the new building, I felt that I had achieved another stage in the overall plan of instantiating a first-class primate research facility at the university—a mini-replication of Harlow's own early career path. The monkeys were housed in rooms attended by trained caretakers who could properly monitor their status and sanitize their cages in machines capable of autoclave-like temperatures. Fresh air was exchanged in the holding rooms at least fifteen times an hour, in stark contrast to the totally stagnant conditions of my sealed concrete room in the chemistry building. The cages were new and shiny, though not much larger than the discarded ones I had brought with me from Wisconsin. I also had an office with hardwood bookshelves and a built-in desk. The entire space was temperature controlled, and the windows were sealed against the blow sand that regularly sifted into my office in the barracks building. I had access to a secretarial staff down the hall, not a quarter-mile walk away.

Shortly after the new building opened, Douglas Ferraro and I received the grant for which we had applied from the National Institutes of Health (NIH) to study the effects of marijuana on learning and memory in rats and monkeys. Douglas was the principal investigator and I was the coinvestigator. Separately, he received a grant to study the effects of cocaine and amphetamine on learning.

In some of those studies monkeys were to be surgically fitted with catheters implanted directly into the jugular vein in their necks and connected to an access fitting protected by a plastic helmet that they wore continuously. With this apparatus, and depending on the design of the experiment, monkeys could receive drug injections timed by the experimenter or self-administered by the monkey by operating a lever according to specific response pattern.

While I could see that this setup would permit the study of interesting relationships, such as the rewarding power of various drugs at different doses, it required surgical skills that I did not have. My grim failures as a rat surgeon in undergraduate school had turned me away from that kind of work. I learned that this jugular infusion methodology could be hard on the monkeys due to abrasions caused by ill-fitting helmets and catheter track infections that were frequently resistant to treatment and required a great deal of handling and physically forced restraint, which was always an ordeal for both human and monkey. I was glad not to be directly responsible for this work but assisted during necessary medical treatments and reviewed the results with interest.

Jeff was anxious for me to meet the people he had been working with at the Lovelace Foundation and to see their testing setup. I was interested in meeting other primate researchers and exploring the possibility of establishing collaborative relations outside the university. The trip, which included introductions and a detailed tour of facilities, turned out to be a disturbing experience, one that required adding thickness to the emotional callus that had been forming on my animal research ambivalence since undergraduate school.

I learned that the rhesus monkeys at the facility were being trained to perform on what is called a delayed match-to-sample task. In this task, monkeys are shown a two-dimensional symbol, such as a green triangle, which is projected for several seconds on a translucent response panel. The symbol then disappears for a prescribed period of time. It reappears amidst an array of other symbols, such as a red circle and a blue square. The monkey's task is to choose the symbol that it first saw by itself.

In the typical training arrangement, a food-deprived monkey

would receive a food reward if it correctly matched the sample, and nothing if it failed. Monkeys would be trained in this way until they reached a high level of accuracy. The training regime that I saw at Lovelace, however, was different: a monkey that failed to make a correct match received an electrical shock instead of losing a food reward. The shock was being used to motivate performance, because once the animals reached the desired performance level they were to be tested while simultaneously being exposed to a source of mixed gamma radiation. Since radiation exposure at the levels used in the studies produces severe nausea, animals trained with food rewards would simply stop responding—the food had lost its desirability. The shock contingencies circumvented this motivational problem and permitted an assessment of whether the radiation negatively affected the short-term memory required to solve the problems. As I recall, the monkeys were exposed to the radiation for periods ranging from several hours to several days.

The researchers explained to me that the research was intended to create a monkey model of a combat pilot flying through a cloud of radiation after delivering atomic bombs on a foreign target. The military question was whether this exposure would debilitate the cognitive ability of the pilots.

My overall impression was that the facilities were excellent, the researchers experienced, and the project extremely well funded. As for the project itself, I had my doubts about whether delayed match-to-sample was a good analogy to the demands on combat pilots, but I could understand the reasoning. I didn't like the fact that the monkeys were tested while immobilized in chairs instead of in chambers where they could move about with some freedom. I knew that chairing was hard on animals, especially for long periods of time. But again, I could see the experimental argument for the procedure. I raised no methodological objection at the time because I had none. If the animals were uncomfortable, it was experimentally necessary.

The tour then proceeded to another set of facilities, where the physiological effects of shock waves on living animals were being studied. Again, the projects that I saw were obviously war related. In ponds filled with water, animals were attached to explosive devices of specific sizes and the devices would then detonated. The nature and extent of the injuries would then be determined.

I heard described but did not see a device called a shock tube. Here animals were placed inside a metal tube on one side of a membrane. On the other side of the membrane, a source of water or air was allowed to build up until it reached the desired pressure. Once that pressure was attained, the membrane would rupture, and a water or air wave would hit the animal with the exact desired force.

There was also a tall tower with platforms set at different heights. Live animals were dropped from those platforms onto what I recall being described as an extremely hard marble floor. I remember the floor because it resembled a black-and-white checkerboard. It was not clear to me whether the animals used in these experiments were anesthetized when they were dropped, but I assumed that they must have been.

During the tour there were stories and jokes about the great sheep barbecues that followed some of the impact and explosive tests. Neither my hosts nor I uttered a single expression of concern for the animals or the appropriateness of the experiments. It seemed so clear to all of us that the work was required to protect national security and the costs to the animals were not a factor. There was simply too much at stake. While I was disturbed by the descriptions of the shock and impact tests, I agreed that they could produce useful data about the resilience of living beings struck by bombs and missiles.

On the day the Psychology Department had scheduled an open house so that the university community and the public could view the new facilities, Chuck, one of our graduate assistants, called me to look at several stump-tailed monkeys that had been recently shipped to the lab for our drug experiments. These monkeys had come from the lab at Holloman Air Force Base, where they were no longer needed. At Holloman the monkeys had been tested in primate chairs, which required that they be restrained at the neck and waist like the monkeys at Lovelace. These monkeys had been trained to walk on a lead from their home cage to the chair, where they climbed up and seated themselves. Each monkey wore a chain collar to which the lead was attached. The collar's links were about the same size as a chain one would use with a dog the size of a Labrador retriever. While the monkey was walking, the lead was pulled through several feet of rigid pipe, which separated the human from the monkey and prevented the monkey from attacking the handler.

While testing these monkeys for tuberculosis, which required that the animals be immobilized while the test agent was injected into an eyelid, Chuck had noticed something quite peculiar about the chain collars and wanted my advice. When I entered the treatment room, one of the new monkeys was lying on the examination table. Chuck motioned to me come to the head of the table. As I watched, he traced the chain collar with his index finger, from the tight loop beneath the chin to the side of the neck. At a point about halfway around the neck the chain disappeared from view and re-emerged on the other side of the neck. I pushed the dark hair away from where the chain should be—and it became clear that the chain had actually grown into the neck of the monkey. Chuck looked at me and said that he had observed the same thing in all the other Holloman monkeys.

I couldn't believe what I was seeing. Clearly the collar had been sized when the monkeys were much younger and smaller, and as they grew the chain became so tight that it had grown into their necks. So the fit of the collar had not been checked for years. And this had been the case in a laboratory that was flush with money, well staffed with veterinarians and trained animal technicians, and supplied with state-of-the-art facilities.

I considered what it must have felt like when the monkey was on the lead and the collar was tugged. While I thought the level of negligence extreme, my expression of outrage was tempered by the recall of my own failures in graduate school and the monkey that I inadvertently starved almost to death. I also thought about the tour I had just had at the Lovelace lab.

Chuck suggested that we remove the chain right then and get this episode behind us. I agreed and assisted as he cut the chain out of the monkey's neck. We did this as visitors walked the halls and even while some looked in to see what was going on in the treatment room. I later learned from the campus veterinarian that recent research indicated that ketamine, the drug we used to anesthetize the monkeys during the chain removal, probably did not provide a fully adequate level of analgesia. Thus we had likely added our own insult to the monkey's long-standing injury. Despite the previous institution's failure and our deficient well-intentioned surgery, though, the monkeys recovered, and their hair growth eventually covered up the long, ugly scar-tissue evidence.

This account makes it clear that major problems in the conduct of animal research are not always caused by intentional and maleficent conduct but may come often from ignorance, carelessness, and error. These kinds of issues are sufficiently common that they're seen as something to be expected, simply the cost of doing business— except that the cost is paid by the animals alone.

∗　∗　∗

As the years 1972 and 1973 progressed, my lab became more and more complete. The caging I had ordered from the same supplier used by the Wisconsin Primate Lab arrived, and the rest of my monkeys were delivered. My work with Douglas Ferraro was going well. I was publishing my own work in good journals, with Harlow as my coauthor. I also found that I loved teaching and received excellent student evaluations. Logan was very happy with both my research and teaching, and he told me so directly. The actual expressions of this satisfaction involved regular invitations to lunch at the Elks lodge just a mile from the department on University Boulevard. At one such lunch he suggested that if I kept up the pace and quality of my work, he would put me up for early promotion and tenure.

As perhaps the clearest evidence that he saw me as destined to become a senior member of the faculty, he said at one of those martini-filled lunches that he had assigned me to teach one of sections of Introductory Psychology the next semester. As I noted earlier, he was highly protective of those Introduction to Psychology assignments and had told me so during my initial interview. From his perspective, Introduction to Psychology was the most important course in the department's undergraduate offerings. It was the opportunity to convince new students of the truly scientific foundation of psychology. It was well known that when Logan taught the course he would sequester himself in his office for several hours prior to each lecture, rehearsing each word, example, and explanation he would use in that day's class.

The story at home was different. Kim was disappointed that I spent such little time at home, and while she was enjoying rebuilding her relationship with her mother, she was lonely and bored. She wanted to start a family, and I was reluctant. While I did not discuss the reason for my reluctance with her, it stemmed from my belief that the excessive number of chronic diseases that had affected my family suggested that my DNA was flawed and would likely create tragedy in any progeny. This was one of the themes that had motivated my interest in biomedical research.

I listened to Kim's concerns but felt that they were somehow not my problem. While I encouraged her to go back to school or find an interesting job, my concern for her was surely perceived as halfhearted. In truth it *was* halfhearted, since most of my heart was promised to my work. What I was doing at the university was my first priority, and her adjustment to the situation I saw as her responsibility. This attitude was based in part on my unquestioned acceptance of the model of professional marriages that I had seen all through school, and not what I had seen at home. There I saw my father as a carefully attentive husband always trying to please my mother and sharing the everyday duties. In undergraduate school, in contrast, Dr. Davidson and Dr. Campbell had often left the lab late at night, seemingly unconcerned about family responsibilities even though each had very young children at home. In graduate school, Harlow and Sackett could be found in their office or lab virtually any time of the night or day. I didn't see their behavior as hostile to their families but as expressions of the urgency and importance that they felt their work commanded.

I wasn't at home because it was so fascinating at the lab. I waited for the results of the experiments on a daily basis. I often sat in the lab, anxious to see the next data point, as the student assistants finished running a monkey or rat. There was definitely a kind of sensuality to the process, immediately gratifying and challenging.

Given my degree of absence, I should not have been surprised to find, when I arrived home from the lab one Halloween night, that Kim had invited a companion to join the household. He was a young black-and-white longhaired German shepherd mix, already named Colin. Kim pointed to where he was quietly sleeping behind the couch. I had seen him running unattended in the neighborhood from time to time and had heard Kim express concern for his welfare. She said that she had found his home and had purchased him from the owner for twenty-five dollars.

I had not lived with a dog since undergraduate school, and then it was for only a short while. I was in favor of the adoption, but there was a problem. We had just signed a new lease that had an explicit "no pets" provision. Yet I could see some joy coming back into Kim's face and realized that we had to make this work. We reasoned that we could make a place for Colin during the day in the small fenced-in spot of ground behind our apartment. I would construct a blind so that he would not be visible from the street when he was outside.

The camouflage plan failed, and within days we received an eviction notice from the landlord. The letter said we had one month to leave the premises and the dog had to be removed immediately. After some cajoling, Kim's mother and stepfather agreed to hold Colin in their backyard until we found a new apartment. It turned out to be a bad time to be looking for an apartment, and we ended up having to buy a small house instead. At the time I had no wish to take on the responsibility of homeownership, as I thought the upkeep requirements would interfere with my work. But winter had already taken firm hold and Colin needed a home, so all objections were overridden.

Colin's presence in the house made an enormous difference in Kim's mood and our relationship. He brought warmth and a focus of care beyond my self-centered attention. My work schedule changed. The three of us began to hike and run together because Colin needed the exercise. We visited the vet whenever there was the slightest hint of a change in behavior that might be a symptom of an ailment. I built him a doghouse, spending hours smoothing all the rough edges and fitting in place the most comfortable mattress I could find.

This attentiveness to the dog's comfort was drastically different from my attitude toward the monkeys and rats at work. There my attention had mostly to do with degrees of deprivation and finding electric shock levels that would efficiently motivate the animals in my learning experiments. While I made certain that the cages were clean and the food fresh, the comfort of the animals was not an issue.

Now I was leaving for work later in the morning and coming home earlier. Kim took up the task of decorating our new house, which, along with caring for Colin, allayed for a while her boredom and resentment.

*　*　*

The final signal of the end of my probationary period as a professor was gaining permission to accept my own graduate students into the program. Up until that point Jeff Sproul had headed a group of highly motivated undergraduates in my lab, while the graduate students that Ferraro and I worked with on our drugs and behavior grant were officially his.

Disappointingly for me, Jeff felt that he didn't have the luxury of attending graduate school and decided that he needed to get a job and get on with his productive life. He chose to pursue a career with the FAA as an air-traffic controller. Harold Pearce, an insightful, obsessively careful experimentalist with incredible organizational skills who had joined my lab, also decided to pass on my invitation to undertake graduate work for similar practical reasons.

My first successful graduate student choices were Robert Frank and Timothy Strongin. Both Tim and Bob had worked in my lab as undergraduates and were familiar with my interests. They were excited, incredibly smart, and hardworking. These two wonderful men, with their ambitious academic plans, quickly made real to me the responsibilities of being an academic mentor. In developing my own supervisory style I naturally drew on the models I had experienced during undergraduate and graduate school. All the faculty members who had been important to me agreed foremost on the importance of nurturing independent experimental thought and being freely available for discussions.

In considering the latter, I remembered meeting Harlow late one night in the Primate Lab as I was passing his office on the way home and he was going to get another cup of coffee. In that awkward moment of a near collision, he asked what I was going to do for my master's thesis. I answered that I was going to test whether early socially isolated and control monkeys differed in their ability to learn that a previously reinforced response would no longer would produce a reward.

He waved me into his office. It was well after midnight. He asked me to repeat my plan. I laid it out in greater detail, and although he looked very tired, he listened very intently. He asked me several pointed questions about the theoretical import, or lack thereof, of the possible outcomes. It was a helpful discussion for me because it forced me to clarify my thinking and expectations in summary form. At the end of the discussion he shrugged his shoulders, raised his hands palms upward, and said, "This is not the kind of experiment I would conduct, but please proceed and good luck." I remember how good it felt to have Harlow express his confidence in me.

I adopted a similar style with Tim and Bob. I pressed them early in our relationship to be creative about their research plans while trying to express the same sense of collegiality that I had felt from my graduate-school mentors. We had discussions at all hours of the day and night. I made certain to spend time in my office when I knew they were in theirs. I encouraged them to read the literature thoroughly and then to let their imaginations go as they considered possible research directions.

As a team, we decided that we would put the emphasis on learning aside for the time being and study the social capabilities of the isolate and control monkeys that had come with me from Wisconsin. As most of the previous research on the effects of early social isolation had examined monkeys soon after their emergence from the rearing conditions, and our monkeys were now sexually mature, we were in a position to study the long-term effects of early social isolation at points that had not yet been examined. We could see whether the isolates' social abilities were still as devastated as they were when they were tested as young monkeys and if there were any changes in the control monkeys. If, as we expected, the isolates were still virtually incapable of any positive social behavior like grooming, play, and reproduction, we could see whether over time they might improve if given more opportunities for social interaction. I was excited that studying social behavior would provide an opportunity for me to apply what I had learned from Sackett about the statistical theory and application of various observational scoring systems, and the possibility of studying factors that improved sociality coincided with Bob and Tim's developing interest in clinical and rehabilitation issues.

There was only problem: our proposed research required that the two groups of monkeys be socially housed for an extended period of time, and we had no facility where this could be done. I went to see Logan to discuss the issue. He asked me to design a modest group-holding facility that I thought would accomplish the goal, estimate the cost, and then get back to him. I think he expected that the assignment would take me most of the year and I would be out of his hair for a while. After all, we had just moved into a new building primarily of his design, and here I was saying that it was not adequate. Bob, Tim, and I sat down, and in short order were had designed a small indoor–outdoor facility made up of two identical housing areas that could fit perfectly on the roof of the new psychology building. I met with the university architects, who were in an unusual lull, and they quickly developed a set of detailed plans for two adjoining cinderblock rooms with adjacent open-air chain-link enclosures. There would be plenty of perches and swings to make use of the vertical space. The architects estimated that the university physical plant could build the facility for about $25,000. Guessing that Logan would suggest funding the construction by securing a small facilities grant from the National Institutes of Health, I began writing up the grant proposal. With the plans and cost estimate in hand, I returned to see Logan. He looked at the plans and said he had money in his budget to cover the construction. The new housing facility was built, wired, heated, air-conditioned, and painted in less than six months.

Instead of having an opening celebration, we immediately began the first phase of the long-term evaluation study by introducing the isolated and socialized monkeys into separate sides of the facility to live together as social groups for a period of fourteen weeks. We had concern over how the introductions would go, given that the monkeys, including the early socialized ones, had spent approximately the last nine years of their lives in small individual cages. There were also some reports in the literature that suggested that isolates became hyper-aggressive when exposed to other monkeys as they aged.[1] Therefore, for each group introduction I arranged for a group of monkey-experienced undergraduates and graduate students and professional animal care technicians to be on hand in case we needed to quickly remove anyone or break up any serious fights. The general plan was to introduce the smallest and most vulnerable animals into the facility first, followed by the larger and potentially more aggressive animals.

When the isolated monkeys were placed together for the first time, they scattered to the far reaches of the facility, keeping as far away from each other as they could. Most appeared frozen with fear and confusion. The levels of stereotyped behavior like rocking and self-clinging increased dramatically and remained that way for several days. We scattered monkey chow and orange and apple sections in both the inside and outside spaces so that the animals would not get overly depleted and dehydrated before finding the food bins and sources of water and learning how to operate the automatic lick tubes.

When it was their turn to be introduced to the group housing, the previously socialized monkeys immediately took up detailed exploration of the facility. They smelled, chewed, licked, and handled every square inch of the structure. While they showed signs of anxiety and stress—vocalizing more than usual, pacing aimlessly, and fighting from time to time—no one was seriously injured, and the distress dissipated rapidly in favor of positive social behavior like nonaggressive chasing, mutual grooming, and multi-animal huddling.

Once the groups stabilized, we began to collect data on behavior and to test how the animals competed to get access to highly preferred food and juice. These competition tests ran the risk of stimulating dangerous conflict but instead generated mostly pushing and shoving. We found that the isolate monkeys were not hyper-aggressive; instead they remained virtually asocial, racked with stereotyped self-directed behavior like self-biting, rocking, and huddling. There was not a hint of sexual behavior on the part of either the males or the females. When competing for highly preferred food items, they quickly developed a hierarchy of access that remained fixed over time.

After three months of group housing, we observed only minor improvements in social behavior among the isolates. They spent more time sitting passively in closer proximity to one another than they had at the start. They also showed minor but increasing levels of active exploration of their living environment. In contrast, the monkeys with early social experience began the process of group formation soon after their introduction, even though they had spent nearly a decade in single cages. They quickly settled on a basic dominance hierarchy and engaged in many kinds of social interactions, including sexual encounters. When tested in the competition tests, they showed, in comparison to the isolates, a less fixed ordering of access and a much higher prevalence of social side encounters.

After the initial fourteen-week period, we decided to mix the isolate and social groups into the same space to see whether that would facilitate the rehabilitation of the isolates by providing them models of relative normalcy. This would leave one of the sides of the facility empty, creating the space we would need to start a permanent stump-tailed macaque colony. It was clear from field and lab studies that stump-tails were a unique group of monkeys, less aggressive toward one another than rhesus monkeys and very intelligent.[2] In my estimation, based on running the two that I had rescued from Wisconsin on learning studies, they were incredibly clever. I thought this type of colony would add an interesting and distinctive dimension to my developing primate lab. While the male and female that I already had were sexually mature, I would need some additional monkeys to create a viable core for a reproducing colony.

I immediately thought of the adult male stump-tail that was quartered at the feed store where I first bought monkey chow when I arrived in 1971. I drove down to the store intending to offer the owner up to $100 in cash for the monkey, assuming that he was still in good health. I was willing to go higher and thought my maximum bid would be $500. On the way, I thought to myself that I hoped the owner had not discovered how rare the monkey was, so that the price would not shoot up.

As I drove into the parking lot, I could not see the monkey in his rickety wood-and-wire house. In the past the monkey would usually emerge from his enclosed compartment when we parked close to his cage. The owner recognized me at once as I entered the store, and we shook hands with smiles and shoulder slaps. I went right to the point and asked how the monkey was doing. He seemed surprised by the question and said that he didn't have him anymore.

"Where did he go?" I asked.

"Oh, he got to be too much trouble," said the owner. "He scratched a customer who was feeding him grapes. I knew I had to get rid of him."

I pressed him further. "So where did he go?"

"Well to tell the truth," he said, "I matched him against my German shepherd about a month ago."

I didn't understand what he meant by "matched," and the question was apparently obvious on my face. He repeated the word again, but this time he mimed the moves of a boxer. "You know, a fight."

I could not speak.

He went on. "He surprised me; he put up a hell of a fight against that dog. He lasted better than fifteen minutes." He was all smiles as he described the details and the final throat-ripping that ended the monkey's life.

I felt a level and quality of nausea mixed with anger — the kind that grips you from your throat to your ribs to the pit of your stomach — that I hadn't felt since the time my father and I were attacked by a street gang in Queens when I was a teenager. During that incident, I was held and forced to watch my father try to defend himself against a group of men who took turns administering the beating. When it finally stopped, a man in the crowd told me to stop shaking and crying and said I should be proud of my father because he had put up "a hell of a fight."

Breaking away from my momentary catatonic state, during which I alternately visualized my father's beating and that beautiful brown monkey with the red blotchy face choking on his own blood, I turned away and left without saying another word. I could see the owner's reflection in the door glass as I reached for the handle. He shrugged and waved his hand at me in disgust.

I had heard that cockfighting and dogfighting were traditional parts of New Mexico culture, but I had not experienced the blood-lust and the indifference to purposely inflicted pain and suffering that motivate that form of "entertainment." Years later, I attended several dogfights in Albuquerque and cockfights at Tommy's Place in Hobbs, New Mexico, with a Humane Society investigator. I was shocked by the level of involvement of young children in the actual fights. I saw preteen children wash their dogs in milk in order to pro-vide evidence to their opponents that there were no toxic substances on their skin and that the fight would be fair. I saw children, both male and female, stroke and hug their roosters as they waited for their turn in the pit, and then throw them half-dead into the garbage if they lost or were fatally wounded.[3]

A few days later I called back to Wisconsin to get the name of a primate dealer from Chris Ripp. As I recall, the information that I received was for a primate importer in New York City. It was incredi-bly simple. I called the number, made an order for three stump-tails, two females and one male, and sent a purchase requisition for some-thing like $200 each. The process was like ordering three stuffed animal toys from the Sears catalog. I was surprised by the low price. There were no questions about the purpose of the order, there was no documentation required to prove that we were knowledgeable about what it took to house and care for a complex and demanding nonhuman primate. If you had the money, you could be the owner of three monkeys captured from the wild in southern Asia.

A month or so later, in early May of 1975, I received a call that three monkeys would be arriving by air in a few days. When I arrived at the airport to pick up the monkeys, I peered into the three wooden travel cages that were lined up on the loading dock. I could see three breathing and alert stump-tailed macaques, two that looked about four or five years of age and a full adult female somewhere between seven and ten years old.

When I got to the university, I transferred each monkey to an individual cage in the holding area in the indoor–outdoor facility so that I could examine them closely for injuries and infections. Fortunately the monkeys were very active and looked healthy. Without hesitation they gobbled the fruit sections we offered them, placing the extra portions into their cheek pouches. The older female had a beautiful, shockingly bright red face, and a brown coat with lighter highlights. Her nipples were elongated, indicating that she had already been a mother. She was immediately named Rose. The male was burly for his age and had a darker-brown coat shading almost to black. He had the very beginning of the longer-hair goatee that would more become prominent as he aged. I decided to name him Israel, or Izzy for short, in honor of Harry Harlow, whose given surname was Israel.

(The story Harry told about his name change was that just before he was about to receive his PhD degree, Lewis Terman, chair of the Psychology Department at Stanford University, told him that he had not been able to place him in a job at either the University of Iowa or the Yerkes Primate Laboratory, primarily because of his Jewish-sounding last name. Even though Harry's religious upbringing was primarily Methodist, Terman strongly recommended that he change his name and offered the suggestion "Harlow." Even though he received a job offer from the University of Wisconsin prior to making the name change legal, Harry went through with the change anyway. While telling this story, he joked that as far as he knew he was the only psychologist who had ever been named by his graduate-school chairman.)

The third monkey was a sexually mature young female with a blondish-red coat like a golden retriever's. Her face was pink with a saddle of dark skin that crossed the bridge of her nose. Her demeanor was calm and friendly. I decided on the name Millie as a tribute to the woman who had been my first teenage love and had recently died in a tragic car crash in New York.

We removed the monkeys from their cages by hand in order to test them for TB. It was incredible how easily they transferred from their cages to the examination table, virtually walking out while we held their hands. They did not threaten or resist except when we pushed them back into their cages. We waited two weeks to make sure that they were negative for TB and were not going to come down with an infectious respiratory disease or dysentery. We then transferred them, along with the two stumps from Wisconsin, Greta and Manny, into the group housing facility. There was no bloodshed, and they quickly began to meld into a functioning social group.

It was clear that the facility had opened up an entirely new research capability for me, and I was thrilled. I imagined many new experiments looking at the dynamics of social behavior as well as opportunities for teaching about primate behavior and observational scoring methods. I had just submitted a grant to study the long-term effects of social isolation, I was admitting good graduate students, and my lab looked like a real primate facility, a far move from my first-year hallway lab in the chemistry building. I corresponded with Jim Sackett and Harry Harlow regularly about possible collaborations. Now the collaborations were to be between their labs and mine, and not just me spending time in their labs like an advanced graduate student. My training with the veterinary staff at Wisconsin had left me capable of dealing with the common diseases and injuries that befall captive monkeys. I successfully treated stress-induced dysenteries and stitched up bite wounds. If I got in over my head, there was Phil Day's Animal Resource Facility on the medical campus. I was surrounded by great students, plenty of monkeys, interesting experimental data, and colleagues who for the most part saw me as a valuable addition to the department. My teaching was going well, and I had reached the finals of the University Teacher of the Year Award (in the end it was won by a senior English professor who was about to retire). By 1976 I had been promoted to associate professor with tenure.

*　*　*

In 1977 I was about to graduate my first MA student. I settled comfortably into my chair, preparing to read the final draft of Bob Frank's thesis, which was titled "Assessment of Long-Term Deficits Produced by Early Total Social Isolation." It was a title and study that fit perfectly into the research lineage of Harry Harlow and Jim Sackett, a lineage that included many outstanding researchers like William Mason, Ernst Hanson, Charley Pratt, and Steve Suomi. I was a part of that lineage, but with this thesis in hand I felt for the first time that I really belonged. I turned first to the acknowledgment section of Bob's thesis, expecting to see an expression of gratitude for my mentoring that would be an additional source of pride and satisfaction. "This thesis is dedicated to the memory of G-44," it began. "She unquestioningly made the greatest and noblest sacrifice a member of *Macaca mulatta* can make; she died while serving science. May she rest in peace."

Once the wisp of a thought that it was a joke disappeared, many things came to mind after I read this dedication. First, Bob seemed to acknowledge that G-44's life had value independent of her participation in science; otherwise the notion of her "sacrifice" made no sense. She was a small, clear-faced, red-blond-colored socially reared female rhesus monkey with a feisty attitude. Had she been miraculously transported back to a forest home in India, she would have likely been capable of participating in the full dimensions of monkey life—having relationships, giving birth and raising offspring, solving problems, avoiding predators, fighting, and dying. Instead her life had been appropriated and consumed for research. Whether the data that we took from her would advance knowledge about behavioral development beyond this particular thesis was not yet known. Nonetheless, the sacrifice was worthy of the loss. In reminding the reader what was at stake in animal research, Bob's dedication was an act of giving back.

Bob had not learned to make this kind of acknowledgment and expression of gratitude from me. While I was not thoughtless about animal welfare, I was by this time hardboiled and goal-oriented. Deep affection and intimate concern for animals was mostly reserved for my dog Colin.

This, in turn, made my thoughts turn to what I had lost during this time of single-minded career development. Kim and I had been finally and formally divorced, and I was living in my lab and a small furnished apartment near campus. That is, there were many nights when I didn't want to return to the empty apartment and instead slept on a cot in the lab. During those nights my sleep was frequently disturbed by the sounds of the singly caged rats scuffling around and gnawing on their hard food blocks and cage bars. However, I had no desire to linger on thoughts of lost animals, lost sleep, and failed marriage. Pride and satisfaction were to be found in research, and I had been very successful at that endeavor.

The whole institutional context of animal research, and indeed that of academia generally, supported and reinforced my single-minded focus on research productivity and career advancement within the university. I received a big ego boost each time a journal published one of my papers, a colleague cited my work, or I was invited to speak at a conference. All of this, in turn, depended on conformance to prevailing attitudes and norms about the use of animals as experimental subjects. For an ambitious researcher in psychology, there was little tolerance for questioning the ethics of animal use, even when it came to experiments as blatantly cruel as those I had seen at Lovelace.

I was in the thrall of a system that could not recognize even the possibility that animal lives mattered except as means to the ends of science. And to reinforce the grip of that system on my conscious mind, I had established entities with their own momentum and self-reinforcing logic: a primate research program and a reproducing colony of stump-tailed macaques. The colony in particular had to be used to justify its existence, and that use entailed appropriation of monkeys' lives. In all these various ways, I was entangled in something for which raising questions about ethics was anathema.

4
Awareness

By three methods we may learn wisdom: First, by reflection, which is noblest; second, by imitation, which is easiest; and third by experience, which is the bitterest.

CONFUCIUS

The realization that something might be wrong with what I was doing with research animals came very slowly and sporadically. From time to time something would happen with the animals, or a colleague would raise a question or make an observation, or I would read something challenging, and the incident would cause me to reflect on the meaning and consequences of my work. During these times, the long-denied self that deeply empathized with animal suffering would find an opening and assert itself, only to be quietly tucked back into a dark corner of my mind by the hard-edged stance I had adopted as an animal researcher—and, perhaps, by the nascent awareness that much of what I was invested in at the time depended on not rocking the boat. The fear of change and the loss of my identity as a Wisconsin-trained primate researcher were major barriers. Although each moment of questioning by itself produced no lasting changes that I was aware of, they had a cumulative impact, gradually wearing down the ability of my rationalizing mind to convince me that all was as it should be.

At the same time, the many years of observing and interacting with monkeys, both casually and in experimental situations, brought an increasing intimacy. The more deeply I came to know the monkeys as monkeys and as individuals, the more difficult it became to maintain the emotional distance supposedly required by the rules of scientific objectivity and to deny the monkeys' obviously complex inner lives and self-conscious awareness of their circumstances. Intimacy and the awareness of vulnerability cannot be separated.

One of the first times I remember questioning in a total, complete thought whether my research was ethically sound was when an undergraduate research assistant named Dan Gorham told me one afternoon in October or November of 1973 that he was no longer able to carry out one of his duties. He came and stood by my open office door, his very flushed face contrasting with his clean white shirt and light-blue lab coat. I invited him to come all the way in, pointing to a chair to the left of my desk. He came in but did not sit down, and instead stood directly in front of me while I sat. The closeness of his position felt intrusive, but it was clear that he was upset about something. "I can't shock those monkeys anymore," he said. "Besides, they scare the hell out of me."

He was referring to an experiment in which monkeys raised in isolation received a brief electrical foot shock to test whether the shock elicited bouts of self-injurious behavior (SIB). It was well known that shocks elicited bouts of aggression in rodents and normal monkeys, so the question was whether it would do the same with SIB in isolate monkeys. My thinking at the time was that if it did, it suggested that the behavior was related to the general category of agonistic behavior. That knowledge in turn might recommend some particular treatment strategies. In fact, brief electrical shocks did indeed elicit immediate bouts of self-mutilation. Dan's distress, as he described it, came from his knowledge that he was the cause of both the pain of the electric shock and the self-punishment that reliably followed the shock. Dan's claim that he was *unable* to deliver the shock suggested that there was some kind of moral barrier in play in Dan's psyche.

Rather than being angry at Dan, I was surprised by the level of sensitivity he was expressing. I say this because the two of us played on the Psychology Department touch football team, and he was not in the least squeamish when it came to delivering hard blocks to opponents.

After I heard him out, my initial impulse was to deliver a sermon about what a research career entailed in terms of tolerating blood and pain in others and to suggest that he look deep inside and evaluate his capacity for such distancing and hardening. I might even have quoted Claude Bernard, the father of physiology, who wrote the following in 1865: "The physiologist is not an ordinary man: he is a scientist, possessed and absorbed by the scientific idea he pursues. He does not hear the cries of the animals, he does not see their flowing blood, he sees nothing but his idea, and is aware of nothing but an organism that conceals from him the problem he is seeking to resolve."[1] I had heard Bernard's position and more modern variations delivered a number of times in both undergraduate and graduate school to students who were wavering in their professional commitment to animal research, and I had fully internalized its proposition that the value of scientific progress trumped other considerations.

I caught myself, however, sensing that such a sermon would not be helpful to Dan at this moment. The resolution Dan and I developed was to reassign him to other experiments that did not involve his handling or shocking monkeys. Another assistant stepped into Dan's slot, and the experiment once again was geared up and running; its results were eventually published after a long delay.[2] The thought that what I had been asking Dan to do might be ethically problematic at its core did not endure.

Another experience that jolted my complacency about animal research occurred during my first extended interaction with S. Bret Snyder, a veterinarian who had joined the Animal Resource Facility at the University of New Mexico and who serviced the main campus, including my laboratory. Phil Day—now a friend thanks to my involvement on the animal care committee—had arranged to tour the chimpanzee colony at Holloman Air Force Base in Alamogordo with Bret and me. Phil had been an attending veterinarian at the chimp facility while he was in the Air Force and had maintained relationships with the staff members there, and Doug Ferraro, my department collaborator, had tested the memory effects of marijuana on a group of chimpanzees from that facility. When we visited one of the main holding areas, I was shocked by the dark maximum-security-prison look of the housing. The chimps were enormous and frightening. Many of them hooted, shrieked, spun in circles, spat at us, and threw feces when we walked close to their cages. Others sat quietly at the front of their cages, their pursed lips protruding through the bars and fingers extended like beggars. I did not know what I'd been expecting to see, but this was not it.

On the tour, Bret and I caught one another's eyes several times, and I could tell he was as appalled as I. If the familiar scene disturbed Phil, he did not show it openly. As a consequence, Bret and I kept our reactions to ourselves, and we did not talk about them on our long drive back to Albuquerque.

A few days later, however, we exchanged face to face our reactions to the chimp visit. The sentiment was that the Holloman chimps had paid, and were paying, a hell of a price for our research on space exploration and now medical advance. Our shared reaction became a foundation for a friendship, one in which we both felt comfortable discussing animal issues openly.

It wasn't long before a professional conflict, however, emerged between Bret and me. It didn't affect our nascent friendship, but it did serve once again to give me pause and to raise some pointed questions. I consulted him about the extent to which I could deprive monkeys of food without endangering their health. The context of my question was that I was developing a memory test for monkeys where I would display to them a sequence of lights on a three-light panel; after a delay, they were to repeat the sequence in order to get access to a small bit of food. The design was based on an electronic toy for children called Simon that challenged players to repeat ever-lengthening sequences of lights and associated tones presented on a circular console.

Unlike children and adult humans, who typically get intensely involved with the game, the monkeys didn't seem much interested in playing. They made very few attempts to follow the sequences, even very short ones. Since most of the monkeys were quite overweight due to the limited activity permitted in their single-cage housing, I thought that they were not sufficiently hungry to value the food rewards.

At the time it was not unusual in many labs to institute a feeding regimen aimed at reducing animals' weight to some percentage of their base weight during the adaptation process. In rats, a common standard was to reduce an adult lab rat to 80 percent of its free-feeding weight prior to testing. This was seen as a normal way to motivate performance. I should point out that in my experience this reduction in base weight was sometimes done quite rapidly, in ten days or less, by installing a drastic deprivation schedule for the rats. In monkeys, the standard regarding deprivation was less well established; I had already reduced them to 10 percent below their base weights, and still they responded erratically if at all. It was clear that the deprivation was having an effect, as they ate ravenously when they were returned to their home cages. I was considering dropping them another 5 percent below base weight and was prepared to go further if need be.

I expected Bret to respond by citing factors like the use of vitamin supplements, control of electrolyte balances, body fat determination, and behavioral markers that could be used to determine whether we had gone too far. While there was some of this on the front side of his response, the core of his reaction came the form of a deeper question. In essence, he asked why I was not wondering about the reasons the monkeys were not responding, apart from their level of food deprivation. "What does it mean," he asked, "that your test does not engage the interest and capabilities of the monkeys? Doesn't their refusal to participate suggest to you that what you are asking them to do, or the way you are asking them to do it, is irrelevant to the monkeys as monkeys?" He went on: "If you have to take a monkey to an extreme level of hunger where it will do anything to get some food, in the end are you still working with a monkey, or have you functionally reduced it to something more primitive, having then defeated the reason that you are using monkeys in the first place?"

He was driving at the idea that I didn't know enough about what kind of cognitive beings monkeys were to develop a memory problem that made sense to them and validly tested what I wanted to test. I was totally unprepared for his questions. I was operating from a perspective of what procedures were "standard" in similar situations, and not questioning my own motivation and knowledge. In other words, research method orthodoxy had replaced curiosity and reflection.

I loved the idea of my test. I could imagine how slick it was going to be to watch the monkeys banging out sequences for a piece of monkey chow. It was going to be so interesting to then test what effect various drugs would have on their performance. Did the drugs reduce or increase the length of a sequence a monkey could learn? Where in the sequence did the errors come? Was it different with different drugs? Did decision making speed up or slow down? I could already see the graphs summarizing the performance of "THE MONKEY" that I would present at a meeting of the American Psychological Association. I could see the interested and envious looks of my rat-runner colleagues in the audience, hear the questions that would follow, imagine the knowledgeable and glib responses I would make. I was now working to bring into existence my dream, and any interference in that actualization had to be neutralized, not necessarily understood.

I was behaving in accordance with the description of the scientist provided by Claude Bernard: single-minded and devoted to my question and my ego needs. It was hard to admit, but I was beginning to see anger developing in my feelings toward the monkeys. While they were the source of discovery, they were also getting in the way of it by not cooperating.

But what surprised me at the moment, given my investment in the study, was how little defensiveness I felt after listening to Bret's criticisms. It wasn't absent by any means, but it did not dominate my reaction. I respected Bret and liked him as a person. He had made a good point, and made it without speaking disdainfully. I felt he offered his questions not as a demand but as a kind of intellectual gift, to be accepted or left unopened. The decision was mine.

After our meeting, I did try to open his gift. What did I really know about the monkeys and how they saw the world? I thought about the design of the Wisconsin General Test Apparatus (WGTA) that Harry Harlow had developed in 1938, and why it worked so well in testing learning in monkeys. In my experience with it, it was the rare monkey that continually balked and refused to perform. Most remained involved and cooperative. After a long talk with Bill Haquist back at the Wisconsin Primate Lab, I had an idea of why this might be the case. Bill pointed out that one reason the WGTA was effective was that the monkey observed the stimuli, made its choice, and received the reward all in exactly the same place. Further, the stimuli were three-dimensional objects that could be handled. In my apparatus, the stimulus observation area and response location were separated from the reward site, and the stimuli were two-dimensional projections of colored light. Another important characteristic of the WGTA was that it inserted a delay between observation of the problem and the monkey's response; the monkey was shown the problem first, and then the stimulus tray was moved within reach of the monkey. Presumably this delay thwarted impulsive choices and forced thoughtful consideration.

I could see some of the possible reasons that the characteristics of the WGTA were so important. Given that wild rhesus monkeys lived in a complex three-dimensional environment, secured food by using their hands, and lived in social groups where there was often intense competition for preferred food, it made sense that these monkeys would be interested in manipulating solid objects and be especially interested if this manipulation had a direct connection to food rewards. My apparatus was not consistent with these essential monkey traits. To make matters worse, correct responses were rewarded with commercially produced "banana" pellets that the monkeys had to learn were food. I was convinced that Bret was right and that I was asking monkeys to do something that didn't make sense to them. This was a humbling blow to my self-assurance, but it did not cause me to alter the overall approach to my research.

Had I considered Bret's question and its implications more deeply, I might have recognized that it contained the germ of a very important idea. I had always assumed, following Harlow, that you studied monkeys either to learn about them as monkeys or to learn about humans. Here Bret was suggesting that this was a false dichotomy, and that to validly study monkeys as human models, one had to first understand monkeys in great depth in their own right. One implication of this contention was that learning about what motivated monkeys, kept them engaged in a task, and attracted their attention—that is, what made them *care* about something—might require delving into monkeys' inner emotional states, intentionality, and self-awareness; in essence, it would require considering the monkeys' point of view. Subconsciously, I probably recognized that putting myself in the monkeys' place and imagining how they experienced the world was dangerous ground.

Soon after this episode, Bret was involved in an event that produced some outrage among some of my colleagues but for me raised some fundamental questions about animal research. As part of his study on the patterns of drug use in monkeys, my grant partner Douglas Ferraro had equipped several monkeys with self-infusion devices with which they could self-administer injectable drugs like amphetamine and cocaine directly into their jugular veins. The setup required that each monkey wear a plastic helmet that served to stabilize the attached catheter and protect it from the monkey's attempts to pull it out or groom the area. While Ferraro himself had little or no surgical experience, one of his graduate students did, and he was made primarily responsible for the implantations and maintenance of the self-infusion devices. While I was interested in the work, I didn't believe I could be of any specific help on the project. Through a period of several months, I noticed that Bret was spending a great deal of time treating the monkeys wearing the self-infusion devices. The facility care technicians explained that he was attending to catheter tract infections, which were becoming increasingly common.

When I arrived at my laboratory one morning, the graduate students were all talking excitedly among themselves, but their voices were at low volume. I approached a small group and asked what was happening. "The vet shut down the self-infusion project," one said. Some of their faces squinted with disbelief. I asked him to repeat what he said so as to make certain that I had heard correctly. "He shut it down!" the student said, more emphatically.

I couldn't believe it myself. I had never heard of an animal research project being closed independently of the wishes of the experimenter. Over the next few days the story spread through the labs, and some of the details began to emerge. Bret had concluded, after months of trying to correct the ways the catheters were implanted and maintained, that the project did not include personnel who had the necessary training and expertise to accomplish these tasks. As a result, the monkeys were often infected and having to be exposed to frequent handling, anesthetic knockdowns, and pain from the infected sites.

The discussion among the students and faculty who were involved in animal research was contentious. Most asked where the "SOB" vet got the authority to make the move he made; they firmly believed that he had crossed beyond his line of responsibility. Underlying this position was the view, still widely held even today, that the responsibility of the professional staff and veterinarians was limited to keeping research going by taking care of the problems identified by the researcher that interfered with progress. They were not supposed to make judgments about the research and who was competent, or not, to carry it out.

There were other, more tentative, voices as well, some of which came from people who were closer to the research in question and had seen the infections and the predicament of the monkeys. These individuals thought the vet had done the right thing, even if they believed that he did not have the legal authority to do it.

The furor reminded me of my past mistakes and failures to properly care for my monkey subjects. I thought again of the Wisconsin monkey I had named Miracle who nearly starved to death due to my carelessness. I wondered what Bret would have done had he been the attending veterinarian at Wisconsin at the time. He was clearly demonstrating that he had limits regarding what level of pain and distress should be tolerated in the name of research, especially when the pain was unnecessary and caused by a lack of skill.

This incident, coming on the heels of Bret's questioning of what I really knew about monkeys as monkeys, gave me real pause. Professor Ferraro was a well-trained and top-notch researcher by any academic measure. If he could make what were apparently serious mistakes, what about myself? There certainly were serious gaps in my knowledge about how to best conduct the research I was directing. What would I have done in my own study in the absence of an activist veterinarian who took the protection of animal welfare in research seriously? I likely would have starved monkeys to the point where the relevance of the obtained data was questionable. Similarly, was the validity of the self-infusion research compromised by the presence of obvious or subclinical infections? How could sick and suffering animals be good research models for anything? I started to think more about what I didn't know—about my specific research focus and about the animals whose lives I had confiscated for science and for the sake of career advancement.

* * *

Another source of questions was my work on the university-wide Animal Care Committee, which I had joined shortly after I arrived in New Mexico. As a member of that committee, I participated in the inspection of animal holding sites around the university several times a year. The committee was made up of several animal-using faculty members from the Medical School, the College of Pharmacy, and the main campus, along with Phil Day, the senior campus veterinarian. Our inspections were limited to the central holding facilities; we scrupulously avoided actual research labs. This was the extent of the oversight authorized by the Laboratory Animal Welfare Act that had been passed in 1966. In the Biology Department we checked the housing of animals like wild rodents, frogs, snakes, jungle fowl, and lizards; in Psychology, rats, mice, cats, pigeons, and monkeys; and in the School of Medicine, rodents, monkeys, and dogs. These inspections focused on whether the holding rooms had a sufficient number of air exchanges per hour to keep the smells down and limit cross-animal infections and whether the animal cages met the modest size requirements called for by the regulations. We checked food storage rooms for cockroaches and invading mice and evidence that the caging was clean.

Our reports, which went to the vice president for research, spoke of odors, bugs, dirt, and cage dimensions; rarely did they mention the animals themselves, and never the experiments in which they participated. Department chairs used our reports as leverage for acquiring additional funds from the central administration for things like automated cage washers, new cages, and newer facilities with larger animal-holding capacities. The reports were not intended to raise questions about the scientific necessity of particular animal experiments or the competence of researchers to conduct them. The sense at the time was that as long as the animals were acquired legally, ate nutritious food, lived in clean cages, and breathed clean air, researchers were adequately meeting their responsibilities.

Even so, many researchers resented the inspections and the tiny glimpses of the research process they provided to outsiders. Some viewed the concern for hygiene as frivolous, wasteful, and even obsessive. A smaller number seemed appreciative of the feedback and worked earnestly to comply with the rules and improve their facilities.

The committee's numbness to the ethical implications of the inspections was aided by the fact that we blindly attributed the animal-care requirements to a governmental bureaucracy bent on expanding its influence. We did not have the benefit of knowing anything about the history of the passage of the Animal Welfare Act (AWA). The impetus for the legislation was the public outrage that greeted news reports that a five-year-old female Dalmatian dog named Pepper had been stolen from the farm of Peter and Julia Lakavage by an unscrupulous animal dealer and sold to the cardiac research unit at the Montefiore Hospital in New York City, where it died during experimental surgery. In a 1966 follow-up story titled "Concentration Camps for Dogs," *Life* magazine depicted the horrid conditions of dogs being held by another animal dealer before they were sold to high-status research labs. The photos depicted starving dogs living in small cages on grounds surrounded by the decaying corpses of other dogs. If we had read the stories and seen the photos, we might have wondered what in the hell was going on. We might have asked why reputable research institutions like Montefiore would do business with dealers that stole pets and housed them in filth with inadequate cover and food. On the other hand, I suspect that even if we had learned about these incidents and their role in creating the required inspections, many of my colleagues and I might have seen the passage of the AWA as an overreaction to the conduct of a few "bad apples."

When the committee toured the dog facilities in the School of Medicine, we encountered mostly mixed-breed dogs, acquired from the city animal shelter, living in single cages stacked two high. While nothing like the noise made by a room full of singly housed rhesus monkeys, the din produced by the dogs' barks and whining was very unpleasant. After a number of visits I stopped touring with the committee when they entered the dog rooms. My excuse was that the noise hurt my ears, but in fact the circumstances of the dogs hurt my heart. I could not stand the sight of the dogs, some with long suture lines on their chests from some experimental procedure, living in their clean but impoverished environments. I couldn't bear to see these dogs climbing and frantically scratching the sides of their cages, their tails wagging wildly, desperately seeking the comfort of a human hand. I knew that the dogs were "well cared for" by the regulatory standards of the day, but this reassurance did nothing to still the sadness, pity, and disgust that I felt when I saw the dogs. I had a dog at home, and living with him gave me a view of what kinds of activities were important to his well-being: exercise and play, resting in different places throughout the day, interacting with people and other dogs. Very little of this experiential life array was available to the laboratory dogs. In addition to these deprivations, they had to contend with pain from experimentally broken limbs, heart operations that had opened their chests, gut-shortening procedures, implanted insulin pumps, and the effects of experimental drugs, to name a few of the possibilities.

No one questioned my excuse as I stood out in the hall while the committee made its inspection of the dog facilities. At a certain level of awareness I knew that my response to the dogs was the front side of a serious question. But I was reluctant—no, embarrassed—to bring it up either to myself or to others.

My strong connection with dogs no doubt played a role in my reaction to one particular experience I had with Bret. It has remained stuck in my mind for more than thirty years. The dog was sleek and black. He was about a year old. He had a mixed pedigree and a heavy genetic dose of Labrador retriever. When I met him, he was wrapped in a blue blanket lying on a stainless-steel examination table. He was in the process of recovering consciousness from the general anesthesia that had been induced earlier in the day as part of an experiment. The specifics of the experiment were unknown to me, save for the fact that an experimental pharmacological agent had been infused into his circulatory system while measurements were made. After completing his recovery, he would be returned to his cage in the vivarium. He had been living in that cage for the past several weeks after having been transferred from the city-run animal shelter.

I was in the recovery room waiting to see Bret. The research technicians had brought the dog into the recovery room and asked if I could keep an eye on him until Bret returned. The technicians knew I was experienced in animal care, so the request was reasonable. I was informed that everything had gone smoothly and recovery had started and would be complete shortly. The dog's breathing was regular, the color of his gums a deep pink (indicating adequate cardiac functioning), and he was beginning to stir.

My job was simple. All I had to do was keep the dog calm and move him to the floor once he had retrieved his senses. He was a handsome dog, and his fur was smooth, glossy, and warm to the touch. His mouth was slightly open, revealing brilliantly white teeth without a hint of yellow tartar. The only visible sign of the morning's experiment was a clean white bandage carefully placed on his right leg where the infusion set had been inserted. He began to raise his head and to draw his legs close to his body. He rolled to the right and started to stand. I supported his balance as he pushed up onto his back legs, then leaned forward. His rear legs quivered slightly as they straightened, lifting the weight of his hindquarters.

As he stood, he readily leaned into my chest, responding to my stroking and firm support. Strangely, however, he did not turn his gaze toward me but instead craned his head forward as if he was seeing something odd and out of place there in front of him. Perhaps he had never been in this room before. I thought that unlikely, as he would have gone through several examinations before being finally assigned to an experimental protocol. Perhaps he was having visual distortions produced by the anesthetic agent or experimental drug.

My ease gave way to clinical concern. Bret was late, and I began to wish he were there. The dog quickly regained his balance and was now standing firmly on his legs. The quiver was gone. He began to strain against my grasp as he continued to stretch his head forward and then from side to side in abrupt darting movements. He made no move to jump toward the floor. Instead he kept up the craning movements. I spoke to him, trying to calm him with my voice as well as with my hands. Since I didn't know his name, I just called him "Pup." Surely he had heard that designation before, offered in kindness by some member of his previous family or the laboratory care staff. In response to my voice, he turned his head toward me, but there was something odd about this move: his gaze did not meet my eyes and seemed only generally oriented in my direction.

At that moment Bret arrived, apologized for being late, and thanked me for watching the dog. "Bret, take a look at him," I said. "There's something odd about him. He seems to be scanning the room in a strange and erratic way." Now we were both holding the dog firmly, looping our arms over his back and pressing the length of his body into our chests from both sides of the narrow examination table. Bret was beginning to examine him with a level of increased concern. Still the dog's head turned and his eyes darted around the room.

Then I realized what was happening. "He's blind, Bret," I said. He was craning his neck in an attempt to see something in a room full of familiar sounds and odors. The lights were out for this young dog, and he was confused and anxious.

The promise of this dog's life had probably begun to erode months earlier when his owners had become negligent or recognized that he did not fit into the household. Perhaps there were insufficient funds to support a fast-growing dog destined to be large. Or perhaps his holes in the backyard were too deep or the puppy damage too extensive. Whatever the reason, the dog ended up at the pound. Before he could be euthanized, a member of a researcher's laboratory staff came to the pound to select some subjects for an experiment. The dog's size met the criteria outlined in the experimental protocol. He may well have been chosen because he approached the front of the cage when the technician walked by. He likely wagged his tail and responded affectionately to the outstretched fingers placed through the wire mesh. These attempts to bond meant that he would be easily managed and handled. A few papers were signed, and the release was officially complete. He belonged to the Health Sciences Center, to be used in the name of science.

As Bret carried "Pup" to another room to test his senses further and call the investigator, I had to wonder about the harm and fear instilled in this simple, affectionate being. Did the data gathered in the experimental transaction somehow balance the costs to this dog? One thing I knew for certain was that the investigator needed to be called to the treatment area immediately to see the harms being lived out in real time on his behalf. Without this experience, how could he or she justify his or her work except by making vague reference to hopes of progress?

After a while Bret returned without Pup. "What did the investigator say to do?" I asked.

"Put him down," he responded. "It has already been done."

* * *

Around 1976 a young woman enrolled in my animal behavior class approached me after class and offered to loan me a copy of a book that she said had disturbed her. It was Peter Singer's book *Animal Liberation*, which had been published in 1975.[3] Although this book is now widely credited with stimulating the rapid growth of the animal protection movement in the United States, I was only vaguely aware of it at the time. When I asked, none of my colleagues had read it, and most had not even heard of it. That fact alone illustrates the extent to which ethics and animal use was left out of our day-to-day considerations. The student had obviously heard me lecture many times about the importance of the work on social isolation in monkeys done by Harry Harlow and others, including me, and was interested in my response to Singer's criticism of that work.

While I had no real interest in reading the book, I did review the second chapter, "Tools for Research, or What the Public Doesn't Know It Is Paying for," out of respect for the student's question. I was surprised to discover that Singer said many things that made sense to me. For example, he said that journal articles underdescribed the painful procedures that take place in experiments because they do not report errors and mistakes such as electric shock or food deprivation going on longer than intended or animals waking up prematurely while in surgery. He also pointed to a study that suggested that less than 25 percent of all animal experiments that were conducted actually found their way into a published manuscript. I knew these facts to be generally correct—actually I thought that 25 percent was much too generous a number.

In his criticism of Harlow, Singer rightly asserted that animals were treated so as to purposely cause psychological damage. He claimed that Harlow knew what the outcomes of his studies would be before he conducted them. He also noted that the language used to describe the experiments revealed an extraordinary level of callousness on the part of the experimenters.

My reaction was immediate. Obviously Singer did not understand the process of science. Yes, it was true that any thoughtful person could guess that a monkey would be fearful upon emergence from social isolation, but they didn't know *how* fearful. What were the ways that the fearfulness would be expressed? Would the fearfulness be modified with time, or would it remain stable or increase? Would the stress affect physiological functioning? And most importantly, could the fearfulness be used to understand and control human fear and anxiety? Only by carrying out the experiments could such questions be answered. And what Singer saw as callousness in the descriptive language I saw as objectivity.

It was also clear to me that Singer was making a judgment that argued that too much pain and distress was being extracted from the animals for the meager benefits that were achieved. I felt very differently most of the time. I believed that any element of discovery was worth the wear and tear on the animals. After all, their value as living beings was enhanced by their participation. They were recruits into a just war against human suffering.

Soon afterward I shared my criticism of the Singer criticism with the student. She listened intently as the two of us sat in my office. At the end of my mini-lecture, I asked if she had any questions. At that point she moved uncomfortably in her seat, reached into her handbag, handed me the university form required for withdrawal from a course, and quietly asked if I would sign it.

My defenses scrambled to protect me. She was rejecting science, not me. Better to let her go. She didn't have the right stuff. I signed the card, and she left without a word.

What was most problematic for me was Singer's reference to studies that were done by Harlow while I was still a student at Wisconsin. These had to do with exposing young monkeys to extended housing in what was called "the pit of despair" (and then renamed the "vertical chamber" due to the negative response of critics). The device, a V-shaped stainless-steel trough with a monkey living in the space where the walls intersected, was intended to create depression in the monkeys so they might serve as models for the human condition. Harlow used his own experiences with severe depression to "intuitively" design the chamber so that it would induce the feeling of being sunk in a well of despair. He was also thinking photographically, as he had recommended that we do when we were his students.

I first saw these chambers when Jon Lewis, another Harlow graduate student, insisted that I go with him to the basement of the lab. He opened the door of a familiar room and pointed at several of the chambers placed around the room, each covered by a mesh grate to keep the monkeys held inside from jumping out. I looked into the chambers and saw the backs of infant monkeys seemingly wedged into their bottoms. Their brown fur was patchy and lacked the typical luster of the young, and the normally clear white skin was red and irritated-looking in spots. Jon could tell that my attention was being pulled toward their appearance. "Urine burns," he said.

We were both troubled by what we saw. The look of the monkeys was quite shocking, even for people like us who had become used to seeing monkeys in total social isolation chambers, restrained in chairs, and recovering from various invasive surgeries. But in a strange way I felt disconnected, and I was not moved to act. Who was I to criticize this work? It would be like telling a soldier who had won the Medal of Honor that he didn't know how to be brave.

A few weeks later I returned to the room to check the condition of the monkeys. This time I noticed that the apparatus had been modified; small tubes had been inserted at the bottom of the V to drain the urine away from the skin of the monkeys. I latched on to this minor change and felt relieved.

Apparently I was not alone in my reluctance to discuss with Harlow an experimental setup that appeared on its face to be too extreme. Debra Blum, in her award-winning book on primate research *The Monkey Wars*, says that both Jim Sackett and William Mason felt that Harlow had gone too far in the vertical chamber experiments and in the grim language he used in describing the apparatus and the behavioral results of the studies. Sackett is quoted as saying that he once took Harlow aside and suggested, "Maybe we should make this work sound a little less depressing." Harlow rejected the suggestion.[4] On the face of Blum's account, Sackett's concern seemed to be with the language and not necessarily the research itself. If any other colleagues have indicated that they more openly challenged the ethics of this and related research practices, I am not aware of it.

*   *   *

One Monday morning in 1978 I arrived at the Psychology Department as usual around 8:00 a.m. As soon as I reached the second-floor stair landing, I knew that something was wrong. I could hear what sounded like monkey screeches and loud human voices. When I opened the door, a monkey ran by in front of me with Tom Hinton, one of my undergraduate students, rushing after it. Tom shouted to me that the monkeys were loose. I hurried to the central animal facility and found that the cages of about twenty monkeys had been opened and the animals were roaming inside the facility and in the outside halls. A yellow chalk message was written on the concrete floor outside the room. It said simply "Listen Torturer: Monkeys Deserve Freedom."

Ector Estrada, the lead technician, handed me a pair of leather gloves and a large catching net, and we proceeded with Tom to herd monkeys back toward the facility and into their cages. It took several hours to get all the animals back in place. We then began the process of evaluating for injuries and then cleaning and stitching up the many bite wounds some of the monkeys had incurred. It was a bloody scene, but luckily the injuries were relatively minor, and all the monkeys recovered.

My reaction was to deride the intelligence of those who thought they were making a strike against cruel science and for monkey freedom. The perpetrators had in fact struck a blow for increased monkey pain and injury. But while all my data and test equipment were unharmed, I felt intimidated. The event had brought home the labs' vulnerability to attack. There were many rants about the animal-rights "crazies" and their antiscientific stupidity around the lab for many weeks following the incident. I was so angry and frightened that it did not occur to me until much later that some of the blood on the floor might have been shed by the "liberators." At the time I didn't consider whether they might have been infected by the dangerous Herpes B virus that some of the monkeys had. I was so focused on myself that I did not think to alert the local emergency rooms of that possibility.

I did wonder whether the student who had brought me Singer's *Animal Liberation* to read might have been involved. Although I knew I could retrieve her name from my class records, I did not pursue that possibility with the campus police. But why not? Was I afraid to look her up? Did I think that the liberators had a point?

The way I framed it for myself at the time was that I didn't have time to waste on the pursuit. The monkey injuries would hold up data collection for a while, and that was the real problem. I had to devote myself to ensuring that no infections developed, because they might further delay starting up again.

The security breach itself was dealt with by installing a better locking system, better key records, and collecting the keys from people no longer working in the animal labs. The break-in was interpreted as a security problem, not an ethical problem that required reflection. The incident had the effect of reinforcing my defenses against the intruding questions about ethics.

Around this same time, I received a call from Dr. Gaynor Wild, a senior professor of biochemistry in the Medical School. He said he understood that I was doing work on the effects of early experience on learning and wondered whether I would be interested in debating him on the nature and nurture of intelligence in front of a class of second-year medical students. He would take the nature side, and I would take the nurture position. The invitation sounded interesting, and the date was open on my calendar. I accepted.

In 1973 I had published a paper with Harlow comparing the intelligence of monkeys raised in an enriched environment with that of monkeys who had been raised in social isolation for the first nine months of life. We reported that the enriched monkeys significantly outperformed the isolates on the complex cognitive tests that we used.[5] This finding had come as a surprise to Harlow; he had long been on record stating that social isolation destroyed social competence but not intellectual abilities. After working on that study, I had kept abreast of the human literature on factors affecting intelligence, so preparation time for the debate would be limited.

The debate began with Professor Wild reviewing the human twin studies that had measured the IQ of identical twins who had been reared together or apart. These data showed a high concordance of IQ scores in both cases. His summary and interpretations of his data were sound.

For my part, I discussed the animal research from rats to my own monkey work showing that manipulating the early rearing environment had a very significant impact on brain growth in rats and learning performance in both rats and monkeys. In my view, these data suggested that intellectual performance was not a fixed trait and was instead quite responsive to developmental opportunities.

During the last section of my presentation I chose to discuss the findings of a recently published book, *The Science and Politics of IQ*, by Leon Kamin, a well-respected Princeton psychologist.[6] Kamin described his experience of examining some of the published work of Sir Cyril Burt, a historical leader in the field of genetics and IQ. Burt had spent his professional life finding and testing identical twins who had been reared apart and publishing the correlations of their IQs. He made three main reports, in 1943, 1955, and 1966, and during this period his sample of twins grew from twenty-two pairs to fifty-three. Burt's core finding was that the high positive correlations between the IQs presented in these three reports were basically identical to the third decimal place. Kamin wrote that when he looked at these reports, he immediately suspected fraud. The possibility of three sets of independent measurements taken over nearly a quarter of a century showing virtually identical mathematical relationships was so vanishingly small that it strained belief. Kamin also presented evidence that some of Burt's coauthors on these papers probably did not exist and that the tests used to measure IQ had questionable validity. I thought that presenting this evidence was important because the expectations of more recent researchers may have been shaped by these so-called foundational findings, and this in turn may have influenced their own data collection and interpretations.

During the Q&A the students asked a host of excellent questions, primarily about the usefulness of early interventions for poor and deprived children, like the Head Start Program created as part of President Lyndon Johnson's Great Society initiative in 1965. I thought the debate had accomplished its educational purpose beautifully.

At the conclusion of the class, feeling exhilarated, I approached my colleague. I praised his presentation, thanked him again for the invitation, and suggested that we do it again. I extended my hand to him. His response was to stand rigidly and glare at me. His right hand remained at his side. As we stood facing each other, his jaw clenched and his face reddened. Had this interaction taken place during my New York youth, my street experience would have warned me that I was about to be physically attacked. My father's advice about not letting your adversary get in the first punch went streaking through my mind. I rejected the covert advice. After all, I was debating a scholar in a classroom where controversy was the nutrient of educational advance.

I stood still in the charged silence for several moments. As I turned away, he spoke out. "I will never invite you again to discuss this issue. You aired dirty laundry in front of these students, which served only to confuse them."

Obviously the "dirty laundry" to which he referred was Leon Kamin's contention that Burt's work was fraudulent. I wanted to respond and repeat my point that, on the contrary, I thought expectations of modern researchers could well have been influenced by Burt's work, and that thinking that data had a purely independent life, uninfluenced by the context of its collection, was naive. But the surprise and tension that I was feeling sapped the fluidity of my thoughts and speech, and I said something that was incomprehensible, even to me. I could see that further discussion was not being invited and would be useless in such a superheated environment, so I just left the auditorium.

Once I returned to the Psychology Department, I sought out my friend Peder Johnson, perhaps the most thoughtful member of the department and an expert on the history of science. I described what had happened and waited for his response. At first he agreed with me that to leave out the Cyril Burt ethics question from a debate about the genetics of intelligence would not have been a credible strategy. After a moment, he added that it is always dangerous to question the ethics of scientists and to suggest in public that not all science is good and worthwhile. It is going to be taken as a personal insult, and the defenses will go up immediately.

At the time, I could not decide what I had learned. Was it that research integrity problems were considered to be so rare that general discussion about their impact on knowledge was unwarranted? Or was it that that I was dealing with an irrational taboo against ethical discussion that had the effect of hiding something that needed the light of consideration?

*   *   *

One Friday morning my friend Bret Snyder, who by now had left the university and was the lead veterinarian at the Rio Grande Zoo, called and invited me to lunch at a favorite Mexican restaurant with the reassuring name of the Sanitary Tortilla Factory. When I arrived, I found Phil Day, the medical campus veterinarian, and Mike Richard, the veterinarian who had replaced Bret at the university, already seated at the table. I learned that this lunch gathering was a weekly "vet seminar," as Phil called it.

I enjoyed the food, the conversation, and the relaxed end-of-the-week mood. Plans for the weekend were discussed along with sick elephants and lab rat respiratory diseases. While I didn't have much to contribute to the clinical discussions, I found it all very interesting.

Later I wondered why I had been invited. While I had good relationships with all these men, particularly with Bret, this really was a group of veterinarians getting together. These men shared an educational history, a professional context, and a sphere of interest. I, on the other hand, was a psychologist and primate researcher, a person often dependent on their expertise to be successful. I wondered how they saw me fitting in.

The Friday vet seminar became a standing invitation, an event that I looked forward to each week. As time went on and the guards dropped, the conversation began to include discussions of problems with people as well as with animals. I heard about zoo administrators who were driven more by a desire for a diverse menagerie than by the responsibility to provide reasonable housing and welfare support for the animals. Bret talked about the "other world" of zoos and said that if you really wanted to determine how good a zoo is in maintaining animal welfare, you had to see the conditions of off-exhibit animals held in the back areas not viewed by the public. I heard about hard-headed researchers who resisted methodological changes that might benefit their animals and their research. I frequently saw my own behavior in such criticisms.

Mike Richard once bemoaned the fact that the rat colony in my department was riddled with chronic respiratory disease (CRD) and that some researchers were resistant to carrying out the steps necessary to clear the colony of this disease. The resistance was based on the traditional belief that this condition was an unfortunate and unavoidable consequence of having a rat colony. Mike reported that some investigators resisted even when he pointed out that the CRD likely affected the behavioral performance of the rats in the studies in which they were subjects.

Similarly, he discussed the difficulty he was having in convincing investigators who performed rat neurosurgery as part of their research to use only aseptic surgical technique. These researchers were so used to doing "bare-handed" surgery and seeing that the animals survived quite well that they considered aseptic technique unnecessary, time-consuming, and expensive. He reported a breakthrough one Friday when he had swabbed and cultured material from beneath the dental-cement skullcaps that positioned brain electrodes on several dead rats and had shown the investigator the bacteria that had grown. That these bugs might have infected the brain and influenced the brain functions under study was beginning, he hoped, to sink in. That these same infections might well have caused the rats to feel ill was an argument that had had no traction by itself.

Eventually I began to realize that my invitations to the vet lunch seminar were due in part to the fact that the vets saw me as an investigator who was on their side. They saw me as a person who at least was capable of becoming troubled by the way animals had to suffer, sometimes unnecessarily, in the service of science. I was not disinterested in the changes they supported: more careful justification of the use of animals, concern for the animals' predicaments, and being willing to sacrifice certain experimental directions when the costs in pain were too high. As Bret put it once, I was starting to develop a "thinking heart."

I was so proud that these men thought of me in this way, yet it conflicted to some extent with my own perception. I still saw myself as a rather rigid animal researcher, mostly willing to accept animal suffering in exchange for progress. It left me wondering just how resistant other researchers were to what seemed to me to be minor moves to improve the welfare of research animals.

*   *   *

Receiving tenure in 1976 had changed subtly the context of my work and made it easier to let my mind wander down new avenues. Since the achievement of tenure means that your job is relatively secure, and the next promotion evaluation is several years away, it provides an opportunity to reflect on the work that led to promotion and whether a course correction is needed. I wanted to think that the work Tim Strongin, Bob Frank, and I were doing had meaningful human clinical implications, but I had questions. We were testing the extent to which older socially isolated monkeys were capable of developing social competencies late in life. We were housing previously isolated monkeys with socialized controls in our group facility in order to see if this experience led to some form of recovery. Harry Harlow, Stephen Suomi, and Melinda Novak had demonstrated that if you exposed young socially isolated monkeys to normally developing infants immediately after they emerged from isolation, they showed remarkable recovery of social functioning.[7]

In addition, Tim was conducting a mini-experiment in which a totally isolated adult female, E-15—a classic isolate marked by bizarre behaviors and no social behavior—was being housed with a three-month-old infant that had been born in our colony. The fact that she appeared to be becoming more willing to allow the infant, Alice, to spend time huddling with and grooming her was very encouraging.[8]

What I began to question was whether my general clinical knowledge and experience were adequate to make a reasoned judgment about the assumptions we were making in our research. For the most part, my assumptions were based on findings from scientists like Harlow and Sackett who seemed very confident that monkeys were in a sense hairy humans with tails. This belief had been reinforced by the fact that it was not at all unusual at the Wisconsin lab for psychiatrists and psychiatric residents to do monkey research. In fact, William T. McKinney, an outstanding research psychiatrist working with monkey models of depression, had been on my PhD examination committee and seemed to take an interest in the clinical implications of my work. I thought that I would have a better perspective if I was more knowledgeable clinically. However, I found that directing a laboratory and having a full teaching and administrative load (I had become assistant department chair) left me with too little energy at the end of the day to accomplish much in the way of disciplined study of clinical issues. I had been attending a day treatment program for seriously impaired psychiatric patients at the School of Medicine that was run by my colleague Karl Koenig, and was finding the patients and his style of intervention to be fascinating. Unfortunately, I didn't have sufficient time to spend with him. Given the fact that I was soon to be eligible for a sabbatical leave, I decided that I would search out opportunities for doing a formal year-long clinical fellowship. Karl was supportive of the idea, and I trusted his perspective.

I wanted to consult both Harlow and Sackett about my plan before I made any specific moves. As it turned out, Harlow, who had by then remarried his first wife, Clara Mears, was due to visit me, so I waited for his arrival so that I could talk to him face to face. When I picked Harry and Clara up at the airport, I was stunned by the change in his appearance. Instead of wearing the baggy suit and overworked demeanor of the past, he was neatly dressed in a pressed suit and fashionable shirt and tie and had a bit of a lilt in his shuffling step even as Clara held his arm in support. Clara was wearing a bright-colored ensemble and a broad smile. They gave the appearance of an advanced middle-aged couple of wealth and position. It felt good to see Harry this way, as he appeared stronger and less frail and vulnerable than he had when I was a graduate student.

As I toured Harry around the University of New Mexico campus the following day, I waited for a quiet moment before disclosing my clinical training plan. His response was immediate. He stopped walking, turned to face me, put his right hand lightly on my shoulder, and said, "John, I can see another man about to be lost to science." His voice was firm, and while there was a trace of a wry smile on his face, he did not seem angry.

I jumped to my own defense, arguing that the move was about improving my perspectives regarding animal models of psychological disorders and not about becoming a clinician per se. He didn't respond in words, but what had been a half-smile broadened a bit. He acted as though he knew something that I had not yet discovered.

Harry's visit was not just a drop-in to discuss research and enjoy the warmth of the New Mexico sun. He was about to retire for real from the University of Wisconsin and was looking for a place where he could be away from those brutal Wisconsin winters and write his memoirs. New Mexico was the first stop on the tour of possibilities. He was in essence determining whether I would like to have him in the same department with me, and whether Frank Logan would welcome his presence.

For my part, I welcomed the possibility. I saw the possibility of an additional string of coauthored publications and long personal discussions about the history of psychology in the twentieth century. Logan, as it turned out, was reluctant to share his small pond with a big fish, even though Harlow required no salary, would support a secretary with his own funds, and wanted only an office and an academic title.

When I called Jim Sackett and described my tentative plan, his response was also rapid but entirely different in tone. "Why don't you come out to Seattle and do a postdoctoral clinical fellowship here at the University of Washington?" he said. "I will check into the possibilities if you are serious."

My response was quick: I knew the internship at the Department of Psychiatry and Behavioral Sciences there was one of the best in the country, and it would be one of my top choices. Later that week he called back and asked me to send a letter of intent to Dr. Nancy M. Robinson, the director of the internship program. In the letter I expanded the argument I had presented to Harlow the day he predicted my loss to science. I pointed out my developing uncertainty about how valid many animal models of mental disorders were and my growing awareness of how little direct clinical experience and wisdom had informed their development. It was my intention to contribute to filling this gap.

Before very long I received a letter of admission and directions that I should arrive in late August in order to participate in orientation and the selection of clinical rotations. This was before I had formally spoken to my chair or submitted the required application for a sabbatical leave to the College of Arts and Sciences. I scurried to back-fill those requirements.

By this time, Frank Logan's alcoholism had gotten so serious that the dean asked the senior faculty to call a meeting to discuss the problem. This was a terrible moment for both Logan and the department. With Frank not present, the dean's concerns were read. They described details of Frank's public behavior and evidence of his personal deterioration. It was obvious that the dean was concerned for Frank personally as well as his impact on the department and university. The discussion naturally turned toward asking Bob Grice, the department's distinguished professor and Frank's closest friend, for his advice. With the room quiet, Grice eyed the group and said, "I don't know what you are all talking about. I don't see any problem."

Oddly, his response provided momentary relief. Maybe the dean was out of touch after all. The absurdity of this position slowly reestablished itself, however, and difficult ethical conclusions were reached. Henry C. Ellis, a prominent human learning and memory researcher in the department, soon replaced Frank as chair. Frank disappeared from the department, and rumor had it he had entered treatment somewhere out of town.

It was fortunate for me that Henry had a goal as chair to improve and expand the clinical area in the department. He raised no objection to my leave request. Instead he expedited the approval process through the department and with the dean. The leave was approved before the end of the spring semester. I packed light, loaded my dog Colin in the car, and headed off to Seattle. In another stroke of good luck, my friend Sharon Landesman, who had been a graduate student with me at Wisconsin but left the program to move to the University of Washington with Sackett, had contacted me and offered the use of a small apartment in her house on Richmond Beach. Everything had fallen in place beautifully.

The clinical assignments that I received were broad and intense. They included working at an in-patient affective disorders ward, a child outpatient clinic, an inpatient child head and spinal cord rehabilitation center, an adult outpatient psychotherapy clinic, a developmental disorder clinic, an adolescent psychotherapy clinic, and a pain management service. The placements provided broad exposure to the varieties of psychopathologies, the treatment modalities used, and the clinical research in progress. The supervision that I received took place both individually and in small groups and was provided by master clinicians who were available around the clock for emergencies.

It was also strongly suggested that all the interns and fellows participate in their own psychotherapy for the length of their stay. A list of clinicians in the greater Seattle area who had agreed to treat interns and fellows at a reduced rate was provided. Based on the recommendation of one of my supervisors, I chose a therapist named Arnold Katz. Initially I saw this relationship as simply providing the perspective of what being a patient felt like, but by the time our work was halted at the end of the fellowship it had become much more—a revealing exploration of my developmental history and many of the moral and ethical decisions I had made throughout my life. With Arnie's help I came to appreciate the extent to which fear had dominated my adjustment to life. My sense of helplessness as my father suffered the ravages of Parkinson's disease had weighed heavily in my interest in neuroscience and research. My foray back into clinical training likewise sprang from those motives to want to learn to effectively benefit others and myself.

At one point in the therapy, I was telling Arnie a story about my grandfather's funeral and the interaction I'd had with him right before he died of a massive stroke. I remember that I was delivering the narrative in a rather matter-of-fact tone. When I looked over at Arnie, I could see him wiping away tears. I realized that he was more in touch with the emotional power of my story than I was. I was describing a psychological weather report while he was feeling the intensity of an emotional storm. With Arnie's help, I was discovering how emotional distancing robs the meaning from your life.

It is hard to communicate how important this fellowship was in helping me mature as a psychologist and as a human being. During my education I had seen much criticism heaped on the psychologist-clinician. Although I had always taken these disparaging views of clinical psychology with a grain of salt, I had few countervailing influences except for Harlow's general respect for that part of the discipline. The fellowship gave me direct experience—as both therapist and patient—of the value of clinical practice and psychotherapy.

One of the central lessons I learned during the fellowship was not to confuse a formal list of symptoms used to define a particular diagnostic category with the nature of the psychopathology experienced by the patient. Symptom lists are concrete markers that identify but do not capture the essence of the vast personal structure of a patient's concerns, hopes, and fears, and the struggle for meaning that surrounds his or her symptoms. In other words, the patient is a whole, unique being with a rich inner emotional life, not an animated machine presenting symptoms. This understanding is fundamental to being an effective and compassionate clinician, but it also has relevance for the experimentalist who wishes to gain insight into human cognition by studying primates. If you take seriously the idea that monkeys make valid human models, then it makes sense to treat them as humanlike in their emotional complexity and self-awareness.

The human, clinical basis for this lesson was solidified in my work with Carolyn, a sixteen-year-old girl who had suffered a devastating brain injury in a car accident on a rural Montana highway. I met Carolyn while I was doing my rotation at what was then known as Children's Orthopedic Hospital, in the unit that worked with children who had experienced a head or spinal cord injury. Carolyn's injury was to her brain stem, the part of the brain that connects to the spinal cord and controls many important functions such as breathing, heart rate, and arousal. She also had an injury to her upper (cervical) spine that left her with only limited ability to control her arms and legs. She could not clearly articulate speech but could make some guttural sounds that did have some value in communication.

On the day we were introduced, Carolyn had just finished a rehabilitation session and was being put back into her bed. She had shoulder-length jet-black hair and shiny pale skin. Her eyes were dark, her facial features were delicate, and her beautifully shaped neck and shoulders were visible through her ill-fitting hospital gown. This natural beauty surrounded a jagged pink tracheotomy scar in her throat, where she had been intubated to keep her alive following the accident.

Once she had been transferred from her wheelchair to the bed, her mother, who was seated at the bedside, straightened the crisp white sheets and tucked them securely around her body. She turned on a small battery-operated tape recorder that was on the night table and adjusted it to what must have been close to its maximum volume. The tape began playing tinny-sounding piano hymn music, which, after a few minutes, faded into the start of an energetic sermon by the evangelist Jimmy Swaggart. There was no way to tell whether Carolyn was listening to or understood the words, which talked of punishment for our sinful lives and the hope offered by Jesus Christ. I winced to myself, imagining Carolyn thinking that her injuries were delivered by a vengeful God as punishment for her sinful nature.

Later, during a treatment team meeting, it was discussed that along with the many disabilities that Carolyn had, she appeared depressed and had begun having intermittent chewing and swallowing difficulties. No one was particularly surprised about the swallowing problems, as they were common with her type of brain injury. The chief psychologist said he had been unable to do any intellectual assessment and was uncertain about the level of her cognitive function; he feared it might be minimal, however, as her history indicated a period of hypoxia of uncertain duration. At the close of the meeting an antidepressant medication was ordered, and I was assigned to aid the speech therapist in developing a program to help Carolyn relearn the proper sequence of chewing and swallowing behaviors. It was felt that my knowledge of learning principles and their clinical applications might be of help.

After about a week the speech therapist and I rolled out our method. All the behaviors required in chewing and swallowing were identified, and the exercises that would be used to practice and integrate them were neatly ordered. We worked with the nurses who fed Carolyn, as they would be responsible for applying the program. They were all experienced in rehabilitation and quickly learned the details. I had high hopes.

But after a couple of weeks of practice, during which Carolyn seemed to work diligently, the intermittent swallowing problems remained. My optimism was shattered. The speech therapist suggested that I go back and observe in more detail the way the nurses executed the practice procedures and what outcomes were produced. Just as before, Carolyn slowly chewed a mouthful of soft food just before she swallowed, the food dribbled out through her partially open mouth onto the bib on her chest. I went back to observe the evening meal, but this time all went well. The next day I observed breakfast and lunch, and hardly any food was swallowed. Perhaps she did not have a consistent appetite due to her depression, and maybe the recently prescribed antidepressant medication would help once its effect kicked in.

Later that same afternoon I found Carolyn alone in her room. I stood by her side as the sun from a rare clear Seattle winter day streamed through her window and across her bed. Since the day we met, I had spent time with her during exercise and eating sessions, but never alone. I spoke out loud, not knowing whether she could understand me. I said that I wished she could just tell me why swallowing was so difficult, particularly at breakfast and lunch.

At that, her eyes locked onto mine, and she slowly raised her right arm. The arm wavered spastically as her hand, shaped as though she were loosely holding a baseball, pointed toward the lamp table at the side of her bed. On the table was a whiteboard, about the size of a large thin coffee-table book, on which the letters of the alphabet, written in red, were arranged.

I retrieved the board and placed it on her lap. I drew up a chair and tilted the board toward her near her right hand. Carolyn started to tap out an answer to the question that I had asked, using the knuckle of her right index finger. It was clear she had understood me completely. With agonizing effort, she continued to dictate her answer as I wrote it down word for word. The process went on for three hours, until her mother returned just before the evening meal. We recommenced the session the next afternoon and then again each afternoon for several additional days until she had written the whole story of what she recalled about her accident and her swallowing problem.

In her urgent and sometimes disconnected message, Carolyn told me that she was able to swallow fairly well when she wanted to.

When the food oozed from her mouth, she was actually trying to spit it at the nurse who was feeding her. As she explained, the nurse that fed her breakfast and lunch most days called her "my grunt girl," and this made her furious. She had no way of telling the nurse how insulted she was and that she wanted her to stop.

All the team's well-meaning theorizing about depression and elements of swallowing behavior had neglected to take into account that Carolyn was a person whose life plans had been destroyed in an instant and was now struggling to regain some control over her body and maintain some sense of dignity. Once the locus of the injury was known and the diagnosis of posttraumatic depression was made, everything had been interpreted from that perspective. Not swallowing was either a neurological deficit or the mechanical impact of a low mood, a framing of the problem that made it impossible to recognize that it was actually the angry response of a person with full awareness and an intact intellect who could not communicate her feelings verbally.

When I told the nurse about Carolyn's revelation, she stood slack-jawed for several moments before sitting down. She looked up at me and said, with an expression of tentative hope, "If this is some kind of sick joke I . . ." and her voice trailed off. I showed her the rich narrative that I had copied down. To her credit, she read it all. She thanked me and said that she had to do something to take care of this. At a team meeting later in the week, I heard that after a heartfelt apology from the nurse, Carolyn's swallowing difficulties had for the most part resolved.

This incident taught me the danger that arises from making assumptions about the intellectual and emotional complexity of a person unable to communicate in typical ways. I realized that I had been willing to accept the premature judgment that Carolyn was only minimally aware of her situation because it eased my perception of just how much she was suffering. Seeing her as a person reduced to basic motivations like hunger and thirst limited the urgency of my responsibilities and made it easier to go home at six o'clock.

The implications for my research on monkeys were clear. As I walked Colin on the shores of Richmond Beach one evening shortly after transcribing Carolyn's message, I recalled the question Paul Kottler had asked when we were studying the reports of the chimpanzee Washoe's command of American Sign Language almost a decade earlier. Paul had wondered what our ethical responsibilities would be if the chimp communicated that he didn't want to participate in the research. At the time I could not take his question seriously. I did not want to consider the extent to which the animals in my experiments might have been suffering in silence and wanted out.

I still had little desire to entertain these thoughts, but the question had become real and legitimate. If I had a monkey alphabet board, what would the monkeys tell me about their life in my lab? I didn't really want to know, but the ethical ground was beginning to shift. The anger over the laboratory break-in had finally dissipated, and I could start to entertain the ethical questions about animal research that had always been in the background.

FIGURE 1 Author at age ten with family dog Prince, 1953.

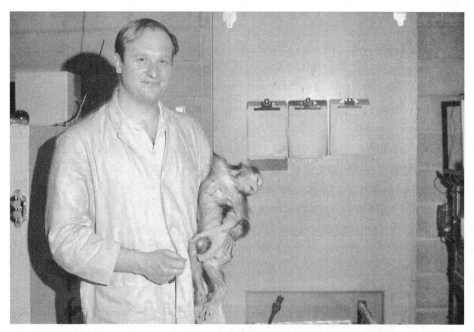

FIGURE 2 Author in graduate school with experimental monkey research subject, 1968.

FIGURE 3 Chamber used to raise infant monkeys in total social isolation.

FIGURE 4 Harry F. Harlow, director of the University of
Wisconsin Primate Laboratories, 1975. Courtesy of the Harlow Primate
Laboratory, University of Wisconsin, Madison.

FIGURE 5 Standard single-cage housing of a rhesus monkey.

FIGURE 6 Rhesus monkey being tested in author's dissertation, 1971.

FIGURE 7 Primate lab construction begins at the University of New Mexico, 1973.

FIGURE 8 Technician Ted Wright and author examine a newly arrived monkey, 1974.

FIGURE 9 A singly caged stump-tailed monkey.

FIGURE 10 Early morning contact between members of a stump-tailed social group.

FIGURE 11 Stump-tails with newborn member of the group.

FIGURE 12 Previously socialized rhesus monkeys living together.

FIGURE 13 Father-infant relationship
in socialized rhesus monkeys.

FIGURE 14 Previously socially deprived rhesus monkeys living together.

FIGURE 15 Bird nest protected by the statue of St. Mary at Georgetown University, 1994.

FIGURE 16 Author with mentor F. Barbara Orlans at the Kennedy Institute of Ethics, 1994.

FIGURE 17 Millie, the last surviving stump-tailed monkey, at forty-two years old, 2014.

FIGURE 18 Author with rescue horse Marigold, 2014. © Jonathan K. Lewis.

5
Realignment

I do not regard myself as what the right-thinking ape would aspire to be.

DANIEL ROBINSON

When I returned to New Mexico in 1978 after my clinical fellowship at the University of Washington, I was not the same person I had been before I left. I knew that I would need to alter the nature of my professional work to fit my new priorities and perceptions, but at the same time I was very reluctant to give up, either publicly or personally, my identity as a primate researcher and director of a primate lab. And even if I had been willing to go that far, radical changes were precluded by the fact that my colleagues, the department, and the university all expected me to pick up exactly where I had left off.

Teaching was one realm in which I could immediately shift my approach to reflect the changes that had occurred inside me. I felt like I had different things to teach and something resembling actual wisdom to share with my students. In both my animal models of human psychopathology seminar and my year-long seminar in the undergraduate honors program, I was determined to take significantly different tacks.

In the animal model course, I took a less forceful stand on the direct relevance of animal studies to clinical matters. My earlier assertion of the primary importance of animal studies had been chastened by the fact that during the entirety of my extensive fellowship I had never once heard a single reference to an animal study as providing a theoretical or treatment insight. This is not to suggest that my supervisors did not believe that animal research contributed to clinical understanding, but it was clear that they did not read or reference the animal literature in their day-to-day clinical teaching or practice, and I had reason to believe they were not atypical. When I would bring up the results of animal model studies during supervision sessions, they listened attentively but then moved on.

My responsibilities in the honors program were primarily about encouraging a small number of select students to design, propose, and then execute research projects of their own creation. If accepted by the faculty, the projects would constitute their undergraduate honors theses. In the past, my approach had been to promote rapid decisions about their selected projects so that they would be able to meet the rigid timeline they faced. Now I found it necessary to talk more and more about what their educational goals were and to reflect on whether the work they were proposing was worth the time of the human subjects that would be recruited or the lives of animals that would be consumed. My interest in promoting self-reflection and compassion had been refined and broadened by my clinical training and experience.

My change in teaching focus was not always greeted with support from students and faculty colleagues. Some graduate students who had committed themselves to a career in animal research chafed at pointed questions about value, usefulness, and the levels of created pain and distress. The undergraduate honors students were more open to and interested in such inquiry, but a few of their faculty mentors told me that my insistence on this type of consideration was interfering with their research productivity. One colleague went so far as to complain that I was holding group therapy sessions during class time instead of encouraging research. I did not completely disagree with his assessment but felt that talking about research goals and how they lined up with personal values was a worthwhile use of class time. His not-so-muted contempt for this "therapy" highlighted a problem that I was becoming more and more sensitive to: the reluctance of many scientists, including myself, to engage in either values clarification or ethical analysis of their work.

The other element of my professional life for which I had a

concrete plan upon return to New Mexico was my clinical work. I knew I wanted to continue my clinical experience and accumulate the hours needed to stand for the New Mexico licensing exam. So, along with my standard teaching duties, I began to see psychotherapy patients at the University Hospital and Student Health Center. I provided my services at no charge in return for formal supervision. This was work for which I had a real passion, and I saw myself as making good progress in mastering the art and science of psychotherapy. Beyond being attentive to behavioral matters relevant to forming a diagnosis, I was learning to "listen" for what was *not* being said by patients during consultations. In the language of psychotherapy, I was becoming attuned to the process of resistance, in which a patient, either intentionally or without plan or awareness, avoids confronting the issues of central concern. As well as symptom-relevant reports, I was attending to the small changes in the pitch or volume of a person's voice, the ways he or she shifted in the chair, the almost imperceptible tear or cough providing indications of a loss of comfort or attention. I began to see how cleverly we avoid the issues that hurt and confuse us, preferring instead to focus on matters that would pass for important but do not upset our contorted level of comfort. I began to understand that successful therapeutic intervention was about the patient's expectations, which in turn were influenced by the therapist's confidence, warmth, and ability to truly understand the patient's predicament. I was learning how to hold a mirror up to patients so that they could see themselves and their choices with better clarity. Yes, therapeutic change required a lot of knowledge about personality theory, neuroscience, and basic research findings, but these factors were useless without a therapist who embodied sensitivity, persistence, and a certain wisdom about life. Therapy needs to engage and help mobilize a patient's intention to overcome the barriers that are interfering with adjustment or transformation. These crucial factors cannot be modeled in animals.

This clinical work consumed a great deal of time and energy, leaving little for research. Unlike the pattern that had prevailed in my earlier career—spending most of my time in the lab—I could frequently be found at the clinic. Frankly, I found this time far more interesting and challenging than time spent in the lab. Harlow, it seemed, was turning out to be half right when he predicted I might become a man lost to science if I pursued clinical training. It was true that the fellowship had changed my perspective on doing science, but it had not altered my commitment to and passion for pursing its goals. I was recognizing that the reality of people's experience—its complexity and the vulnerability and suffering it involved—called for a different science from what I had been involved in. It certainly called for an understanding of developmental processes, cognition, memory mechanisms, and the characteristics of psychopathology. But it also called for appreciating the ways in which these basic processes were affected by a patient's trust, self-confidence, comfort with intimacy, and acceptance of the unpredictable and often cruel contingencies of life.

In 1979, after having been licensed as a clinical psychologist for about a year and starting to supervise clinical graduate students, I received a surprising phone call from the office of the newly elected governor Bruce King. The representative from the division of Boards and Commissions indicated that Governor King wanted to appoint me to the New Mexico Board of Psychologist Examiners, the body responsible for licensing clinical psychologists and investigating complaints about their professional conduct. They asked me to take a week to consider the offer.

While I had voted for Bruce King during the election, I had not contributed financially to his campaign, which I had assumed was necessary to be considered for appointments to state boards and commissions. I soon realized that the appointment had been facilitated by my friends Karl and Frances Koenig, influential psychologists who were also politically well connected. I was moved by their confidence and quickly accepted the appointment despite recommendations to the contrary by several of my senior colleagues. They argued that the work would consume a great deal of time, would not be seen as valuable when I went up for promotion to full professor, and would surely reduce my research productivity. I was further surprised when Governor King called me personally to thank me for accepting.

My time on the board (which ended up being twelve years) changed my professional life by requiring that I focus my attention on the lapses in professional conduct that constantly came to the attention of the board. As a member and chair, I had the opportunity to listen to questions and complaints from the public about the treatment they received from psychologists whom they trusted to care for them. I heard astounding stories from the beginning. My professional life was becoming full of the recognition of unethical professional conduct.

Early in my tenure on the board, I received a message from the board secretary asking me to respond to a call she had received that morning. The woman who called was not making a complaint but rather had some general questions about appropriate treatment strategies for anxiety disorders. I returned the call, and the woman — I'll call her Darla — asked if she could meet with me face to face instead of talking on the phone, because she wanted to ensure that the conversation would not be overheard. It was clear that she was seeking more than just information.

We met at my university office several days later. Darla described herself as a small business owner with a husband and two children. She was college educated and spoke with a sophisticated but not pretentious vocabulary. She went on to say that her business had been experiencing difficulties, and as a result her level of anxiety had increased significantly and her sleep had been disturbed. She had also begun smoking again, which upset her greatly as she had worked hard to quit a long-entrenched habit a couple of years earlier.

She had sought help from a psychologist who advertised his expertise in treating smoking addiction, assuming that he could also help her with her anxiety and sleep difficulties. After a couple of consultations, the psychologist suggested that she learn relaxation methods. She went on to say that during a recent session the psychologist was instructing her in the use of diaphragmatic breathing when he moved his chair next to her and asked her to close her eyes. He spoke softly and instructed her not to resist as he slid his hand down the front of her blouse and rested his hand between her breasts. With his hands so located, he pressed and lifted as she breathed in, and pressed gently when she exhaled. This went on for several minutes, during which time she felt the fingers of his hand spread out and contact a larger area of her breasts. She was made very uncomfortable by the maneuver but did not object as she had been told.

I wanted to interrupt her and ask why she allowed this man to continue with what was obviously unnecessary groping, but I stayed quiet. At this point she turned to me and said that her question for me was whether the psychologist was behaving in appropriate ways in teaching her how to use her breath to relax.

When I first met her, I would not have thought that this bright, educated, capable, and sophisticated person could be so easily manipulated. Up until that moment I thought I fully appreciated the problem of vulnerability and its impact in clinical situations, but clearly I did not. Defining *vulnerable* as merely weakened, confused, and in need did not capture the process. (Actually, the Latin root of the word does capture the depth, as it refers to a wounded soldier lying defenseless on a battlefield, completely at the mercy of the able-bodied combatants.) In the time that followed my emphatic "no" and description of the process of filing a formal complaint, I realized how Darla's anxiety disorder had weakened her competence, autonomy, and sense of trust in her own intuitions. It was also clear that her ingrained deferential attitude toward doctors had joined with her anxiety to further deepen her vulnerability.

During the formal hearing that took place about eight months later, the psychologist aggressively defended his touching as accepted clinical practice and even presented an expert witness who supported his claim. After the assistant attorney general had finished his cross-examination of the psychologist, I, as the hearing officer, also had a turn. Seeking to probe the psychologist's understanding of the patient's perspective on the touching, I asked him at one point what he thought was going through the patient's mind as he touched her breasts. Without hesitation, he turned in the witness chair to face me directly, looked at me with a knowing man-to-man leer, and said, while almost licking his lips, "She was loving it."

His attorney jumped to his feet, shouting, "Move to strike, move to strike!" But it was too late; this "professional" had revealed his predator's heart.

During the years that followed I encountered many problematic cases that illustrated various levels of ethical disability. One psychologist sold a house to a current patient; another wrote a book about a particular patient's treatment without developing a clear consent agreement; several used their authority to insist on an excessive frequency of treatment sessions for overly dependent patients; a number terminated treatment prematurely to pursue a personal relationship with a patient; and one convinced a patient that oral sex with him was an essential part of the overall treatment regimen. Along with these dramatic instances were many other examples of ethically rudderless behavior. Many of these experiences shocked me, especially when I knew the psychologists professionally or personally. Most of the time I could not understand how their unethical conduct mapped onto the person I thought I knew.

I eventually came to believe that some people who find themselves in the presence of profound vulnerability are not moved to demonstrate care and protection but instead see an opportunity to advance the satisfaction of their own desires. I wondered how the worst of these actors could have slipped through graduate clinical education undetected. This taking-advantage dynamic, though insidious, would not be hard to see with the close faculty contact required by graduate training. In fact, the investigations we did in preparation for prosecution frequently revealed that questions about the accused person's character had indeed been raised by teachers and supervisors during training. When these individuals were questioned further, they said that they were perplexed about how to handle their concerns, and most chose in the end to remain mute and pass the problem on to the next educational level. This behavior seemed to stem from a mixture of cowardice, a sincere uncertainty about their own ethical judgment, and a fantasy that the licensing process was capable of making such crucial determinations or that the person would straighten out on his or her own.

There were several cases in my own department of giving passes to ethically challenged clinical students. Typically, a faculty member would openly express concerns about a student's ethical judgment because he or she behaved, for example, as if respect for the patient–therapist boundary was an archaic and optional rule. If other faculty members shared the concern, and if the student expressed resistance to an amelioration plan, it would be recommended that the student be terminated from the clinical program. However, the student would then invariably be offered the option of continuing graduate work in the department's experimental research track. If the new research focus was to involve animals, so much the better. The way in which animals — the paradigm example of vulnerability — might be treated by these failed clinicians was hardly given a second thought.

These decisions exemplified the incredible, mostly unspoken belief that the research context does not present the same types of potentially harmful investigator–subject conflicts and opportunities for exploitation and therefore does not require experimentalists to have characters as ethically sound as those of people working in a clinical area with clearly defined "patients." Supporting this belief is the impression that in research there are many imposed rules and regulations and oversight committees that serve to keep the potentially wayward researcher in line. The position is naive and rife with self-imposed blindness. It fails to recognize that research, like clinical work, typically takes place behind closed doors, away from outside scrutiny, giving ample opportunity for asserting a self-interested advantage if such a desire is present. How a professional in a position of authority responds to the presence of vulnerability — in the form of a compromised human patient, a college sophomore required to earn class points by "volunteering" to participate in faculty research projects, or a totally dependent animal held in a lab — is the issue. Does the individual recognize the existence of vulnerability and its existential meaning? Does that person bring to interactions with the vulnerable a desire to be guided by a moral identity as a compassionate and ethical person who is aware of the potential for exploitation and unnecessary harm and the need to tame one's self-interest?

My work with the board of examiners and psychotherapy patients was fulfilling and consistent with my changed values. It felt right, in spite of being emotionally draining. My research, however, was a different story. I was no longer completely sanguine about the usefulness of primate models, and I felt increasingly conflicted about doing animal research. Nevertheless, while a major change in ethical focus smoldered, I worked to maintain the primate-researcher identity that I had built up over the years. I gave invited talks and published a number of papers that still identified me as a primate researcher concerned with the long-term effects of early social isolation on both social and learning ability. After Harry Harlow died in 1981, I began writing an article about him that was clearly a tribute and offered no hint of the break from his scientific lineage that I was in the process of contemplating. In 1982 I debated Steve Suomi, my Wisconsin colleague and friend, not about whether primate models were clinically valuable, but on whether there was evidence that reversing the disabilities induced by social isolation was possible in older animals. These activities were more for show than genuine reflections of personal passion and commitment. The truth was that my professional life was dominated by my work with the psychologist licensing board and my expanded clinical responsibilities. Along with my teaching, these duties adequately filled the space in terms of professional identity—and these circumstances proved remarkably stable for the first half of the 1980s.

One happy consequence of the end of my obsession with animal research was an opening up of emotional space and time for more of a social life outside of work. I also had more time for romantic involvement and relationship, and in this realm I focused my efforts on Charlene McIver, a woman I had known since my earliest days in Albuquerque. A single mother with two children, she had become a graduate student in the clinical psychology program and had focused her dissertation research on the psychological consequences of cardiac bypass surgery in middle-aged military veterans. After she finished her clinical training in San Diego, I convinced her to return to Albuquerque and be my wife. Once we were married in 1981, I was suddenly not only a husband again but also a stepfather to two delightful children, nine-year-old Jeremy (or Jay) and six-year-old Katie. At one point I invited Katie's school class to visit the monkey lab and observe the monkeys in the social groups. We also examined an anesthetized monkey up close so they could see for themselves the humanlike characteristics: the shape of the hands and feet, the presence of fingernails, the forward-set eyes. The children loved the experience, and Katie became proud to say that I was her dad.

*   *   *

In the middle of the 1980s several events combined to shake up my professional life and reinvigorate my ethical reexamination process, which had in recent years been stunted by my own psychological resistance. One of the most consequential of these events was the passage, in 1985, of substantial amendments to the Animal Welfare Act, which had originally been enacted in 1966. These amendments abruptly altered the landscape of animal research, in part because they required that all facilities and institutions using animals for research create formal Institutional Animal Care and Use Committees (IACUCs) with considerable authority to regulate animal use. While the original AWA legislation had helped to ensure appropriate means of animal acquisition, better facility hygiene, more nutritious food, and medical care for sick and injured research animals, it did not permit inspection committee members to enter laboratories and see how animals were actually treated during the process of experimentation. The 1985 amendments changed this and more. IACUCs were empowered to directly inspect laboratories, and researchers were required to submit protocols to the IACUC for review and approval prior to initiating an experiment. This review process involved providing evidence that the work was not unnecessarily redundant, had been planned in such a way as to reduce the number of animals to the minimum necessary, and minimized the amount of pain required by the experimental question, and that serious consideration had been given to the use of nonanimal alternatives. These changes opened the way for members of IACUCs to get a better sense of how investigators designed their research, how they actually treated animals while conducting research, and how the animals fared in the process.

At the University of New Mexico, a newly created IACUC replaced the former Animal Care Committee, and as a long-standing member of the latter I was appointed to serve on the IACUC. I was subsequently elected chair of the committee. In this capacity, I was pleased to see that all sorts of investigator attitudes and practices broke to the surface thanks to the changes in the law. Sometimes the revelations were reassuring pictures of competence and caring, but many others were halls of horrors. Some investigators were deeply hostile about having to provide justification for their use of animals; others complied but provided improbably low estimates of the level of pain and distress experienced by animals during experiments. The committee encountered many examples of researchers deluding themselves about the level of pain their animal subjects would experience and a few instances of causing unnecessary pain and suffering because of ignorance of animals' physiologies or anatomies or proper surgical techniques. There was evidence that some research programs were driven by nothing deeper than inertia and the need to be doing just something to maintain credibility.

When Frank Logan, the former chair of the department of psychology and a well-respected internationally known learning researcher, submitted a research protocol that involved studying the behavioral adjustment of rats exposed to different patterns of food reward tested in a straight maze, his justification was simply that he had been doing "similar experiments for decades." When the committee followed up and asked the obvious question of what was new in the current proposal, he was insulted and angry. He explained to me that he felt that his reputation should have provided sufficient justification and he had no intention of providing anything else to the committee. "Pearls before swine," he muttered. I tried to cajole Frank by drawing his attention to the fact that when the Royal Society of London began its regular discussions of experimentally produced data beginning in 1660, their motto was *Nullius in verba*, meaning "on the authority of no one." When the committee rejected his reputation-based claim of competence as a sufficient justification for his rat experiments, he withdrew his proposal. Several months later he directed his assistants to euthanize all his rats and dismantle the testing equipment and dispose of it. When I asked him why, he said that apart from the personal insult, he had no wish to talk about affective states in rats and how he intended to take them into consideration. "I refuse to do it," he said.

The animal research world had taken a distasteful turn for Frank Logan and many others. Authority and reputation were no longer sufficient to garner approval for research activities. Instead, judgments of experimental acceptability were based—in theory at least—on the nature of the experimental plan, purpose, and methodology. It was a process that asked the researcher to consider the costs in harm to the animals involved and delineate how those harms could be eliminated or at least limited.

While Logan's response to the prereview and approval requirement was extreme and atypical, the sentiment behind it—"who are you to judge my scientific purpose and methods?"—was not. As chair of the IACUC, I was beginning to have many uncomfortable interactions with investigators who, like Logan, openly resented committee questions. I was initially sympathetic to the natural resistance to an additional task, but the level of hostility and condescension I encountered was at times incredible. Some researchers accused the committee of having no members intellectually capable of understanding the purpose of their methods and design. In such cases the committee had to insistently point out that response to the inquiries was not optional and that continued failure to cooperate with the committee would result in the termination of research privileges.

Because of my developing questions about the ethics of animal research and my own involvement in it, the most interesting aspect of my work on the IACUC was what became visible now that the laboratory and classroom doors were partially opened by the new regulations. At one of its early meetings, the committee was deliberating about a teaching proposal that required the use of ten "random source dogs." They would be used in a laboratory exercise for medical students who were studying cardiac physiology to illustrate some of the key reflexes of that system. In their proposal, the professors argued that the exercise was vital for the students because it would provide them much needed hands-on experiences that went beyond book study of the heart and its pathologies. The students would be required to open the chests of the dogs and attach the various recording instruments themselves. They would, the protocol argued, learn basic surgical skills, become familiar with some of the experimental techniques of the area, and, in the words of the instructors, "get the feeling of working with real flesh." The proposal stated that there was no adequate alternative available to provide these important educational experiences. At the conclusion of the class, the dogs would not be allowed to regain consciousness and would be euthanized. These arguments were persuasive to the majority of the committee, and the protocol was passed. Several members requested to observe the laboratory exercise and were reluctantly granted permission to do so. The reluctance of the instructors appeared to stem from their feeling that observation by committee members was totally unnecessary and might interfere with the class. I was one of the observers.

When the committee representatives arrived at the classroom, the dogs had been anesthetized and arranged on their backs on narrow stainless-steel tables. Cords pulled the legs away from their bodies, leaving them in splayed positions. The students approached their assigned tables and waited for instructions. The professor in charge, a cardiac physiologist, told the students to shave the chests of their dogs in preparation for the incision. Once the group had completed this task, they were further instructed to take the "cauterizing unit" and begin to cut through the flesh covering the midline of the chest. The protocol had specified the use of "surgical cautery units," devices that both cut the flesh and seal the severed blood vessels with high heat. As the students awkwardly began, I looked carefully at the devices they were using. They were not surgical cautery units at all but common gun-gripped soldering irons, the kind found in countless garage workshops. Since the soldering irons lacked sharp cutting edges, the students were reduced to burning and tearing through the dogs' chest tissue. Some of the students seemed tentative and dismayed, while others hunched diligently over their dogs, directing strong raking movements back and forth over the center of the chest. Thin swirls of smoke rose from the incision sites. Around the room some students stood still, resting their fatigued hands and arms. The room quickly filled with the acrid smell of burning fur and flesh.

One of the medical students, who seemed particularly hesitant about the exercise, asked the professor what anesthetic was being used on the dogs. It was obviously a question about whether the dogs could feel any pain from these crude surgical procedures. He responded that the agent was alpha-chloralose. When the student said that he had never heard of that anesthetic, the professor explained that it was not used in clinical medicine but had been chosen because it did not interfere with the cardiac reflexes that were to be highlighted in the exercises. The student began to ask another question about the level of analgesia provided by the agent when the professor interrupted him and told him to get on with the dissection and measurements that were the focus of the class. It seemed to me that the professor was so certain of the usefulness of the exercise that he no longer saw what was actually taking place. The class exercise did not really provide the students an opportunity to experience the "feel of flesh" from a surgical perspective, but it did desensitize them to the presence of blood and gore.

After the class I looked up the characteristics of alpha-chloralose and was surprised to find evidence of a controversy about whether it provided adequate analgesia for deep dissections like the cracking of the chest of a dog. I read recommendations not to induce anesthesia with alpha-chloralose for invasive surgery without providing adequate doses of strong analgesics like morphine.[1] The thought of those dogs being chemically restrained while experiencing incredible pain made me shiver.

A week or so later I brought this article to the attention of the professor. He interrupted me before I was finished and assured me that he was the expert on the relevant characteristics of alpha-chloralose and whether its use was justified. It was as though the question of pain was irrelevant as long as the purpose of the exercise was physiologically sound. The professor's arrogance and indifference were overwhelming. (In recent years nonanimal alternatives for teaching this material have been widely accepted across some 125 US medical schools.)[2]

I could list many more examples of naive ignorance, careless determination, self-serving evaluations of experiment-produced pain and distress, emphasis on convenience, and ego-driven defiance. However, the point I want to make, besides noting the existence of these antiresearch attitudes, is that these experiences with the IACUC intensified my questions about the animal research enterprise in general and my own animal use in particular. I discussed these encounters with Charlene, herself a clinical psychologist with considerable human research experience, many evenings after work. She had an intuitive grasp of the ethical issues that emerged from my descriptions. When I tested the idea of ending my involvement in animal research, Charlene encouraged me—no, insisted—that I keep this self-examination going.

* * *

One day in 1985—I think it was after the formal organization of the IACUC on campus—I decided on the spur of the moment to attend a colloquium in the Department of Biology by the eminent researcher Donald Griffin of Rockefeller University. Griffin, a highly respected zoologist and member of the prestigious National Academy of Science, was giving a talk on animal consciousness. He had published a small book in 1976 titled *The Question of Animal Awareness*[3] in which he challenged animal behaviorists to shake loose from the tradition of excluding animals from the select group of organisms (i.e., humans) possessing consciousness and proposed the need for a field he called "cognitive ethology."

His point during the lecture was that careful observation of animals in nature shows many species capable of enormous versatility of behavior and complex patterns of communication. As an example, he discussed the fieldwork of Cheney and Seyfarth,[4] which showed that East African vervet monkeys were capable of communicating to the group the presence of three different predators—eagles, leopards, and pythons—with three distinct vocalizations and that the calls produced three different behaviors in group members: looking up for eagles, running into the trees for leopards, and looking down for snakes. Griffin argued that if these adjustments and communications had been observed in a new group of humans, we would likely acknowledge that a process of thinking and intentional symbolic communication was taking place. His point was simply that ethologists (and comparative psychologists) needed to begin designing experiments capable of investigating whether high-level mental processes might be present in animals.

The idea of animal thinking was very familiar to me because of working with Harry Harlow, who showed that monkeys demonstrated complex intelligence. However, where Harlow had seen in monkeys' cognitive abilities only the relevance to competing theories of learning and failed to see the ethical implications of his findings, Griffin seemed open to considering the possibility that animal consciousness implied a certain moral standing. In response to several questions, he conceded that the presence of consciousness might have ethical implications but demurred from explicitly expanding on the topic. He did say that he rejected the assertion that research that found animals to be possessors of complex mental capabilities somehow threatened the foundation of human rights and the special treatment demanded by their unique capabilities.

What stunned me was the reaction of many of the biology faculty and graduate students to his presentation in general. To say only that they rejected his mental capability proposal is to leave out the rudeness, condescension, and raw hostility that were openly expressed. Many attendees left the auditorium early, making no attempt to do so quietly, and others waved their hands in disgust. Intensely questioning the soundness of a scientist's proposal is a common and acceptable reaction and the purpose of giving presentations and writing papers. Debate is the heart of scientific advancement. But the reactions I saw went far beyond that and lacked any respect for the opposing viewpoint. The respondents didn't just say, "I disagree and here is why"; they seemed to also say, "How dare you!"

Griffin's presentation clearly threatened many of his colleagues because he called for a substantial departure from a set of premises and methods that formed the footings of their teaching and research. He was asking animal behaviorists to get past the emphasis on clearly observable behavior and to create methods that investigated the presence and evolutionary adaptiveness of consciousness and thought. The reaction was a clear example of the conforming power of paradigms in science: a set of ideas and assumptions becomes so taken for granted that most practitioners, in their comfort and complacency, reject any call to scientifically consider whether new paths need to be explored. When I had read Griffin's book years before, my reactions had been much like those I was seeing in the audience that day. But now I was willing to allow his messages to enter my internal debate.

I had continued to interact regularly with Bret Snyder, the veterinarian who had questioned my food-deprivation methods early in my career, even after he left the university to become the lead veterinarian at the Rio Grande Zoo in Albuquerque. We met at least weekly for coffee and games of one-on-one basketball. These hours together were full of discussions about animals, their incredible capacities, the confinement-induced diseases he saw at the zoo, the experimental use of animals, and animals' ethical place in a human world. He was still asking hard questions about my animal work. I consistently left our discussions smarter, enriched by friendship, and having been challenged to see things from different perspectives. During one of these discussions, Bret asked why I was not spending more time in the lab. I considered his question very seriously. I had clinical responsibilities that took up a great deal of time, but that wasn't the total story. I had gradually shifted my responsibilities for the management of the research to an incredibly capable graduate student, Alan Beauchamp, and I had complete confidence in the abilities of my two animal technicians, Ector Estrada and Gilbert Borunda, to take care of the animals. Isn't this the way it is supposed to evolve? The head of a lab directs and delegates, and need not be involved in the hands-on work once a responsible and competent set of technicians and students has been assembled. Wasn't this a sign of a scientific maturation and success? Yes, maybe for someone else, but not for me. The problem was that I really didn't want to be in the lab anymore.

A month earlier, I had been in one of the animal colony rooms checking out a lameness that had developed in one of the monkeys. My attention was drawn to G-49, or Moose, one of the monkeys who had come with me from the University of Wisconsin. She was running from one side of her cage to the other, freezing momentarily when she got to a wall that was made of solid stainless steel. I couldn't understand why she was holding still at that location. I moved slowly to one side so I could see her more clearly while she faced the cage wall. After a few minutes of observation, I could see what she was doing. The wall, which was removable, had one missing bolt, and this left a small hole. If Moose held her head at a very specific angle, she could see through the hole a bit of several monkeys normally occluded completely from her view. She would look at the monkeys through the hole, back away from the wall, make a social signal of some kind, and then return to the hole to see the response. Of course the other monkeys had no idea that she was sending any social messages through that half-inch hole.

In that moment I was overwhelmed with the realization of the dreadful life she was leading and had been for years. Her soft, strong body and curious individuality were encased in a hard metal box, and yet she struggled against its restrictions, working to keep alive some semblance of a monkey social life. Her attempts to project her facial expressions and body language through a hole at a few animals similarly restricted were excruciatingly pitiful. In those moments a connection was established between Moose and me—an emotional connection forged by simple, uncomplicated concern for her welfare. That connection had been there briefly when I first met her as a graduate student but had soon been buried beneath the desire to use her life for discovery.

I started to look at the other animals in the room, and the circumstances of their crushed and distorted lives rushed at me. There was Charley, whose Plexiglas helmet, which kept in place a catheter, was causing abrasions on his neck and making it difficult for him to see because it tended to slip forward over his eyes. The female E-38 paced in clockwise circles, executing the remnant of a behavior related to her desire to exercise and explore. J-90, a large rhesus male who lived at end of the lower tier of cages where there was very little light, looked as if he was staring out of a dark solitary confinement cell.

I finished the examination of the lame monkey, wrote a note to the vet, and went back to my office as quickly as I could. A clinical student arriving for a supervisory session saved me from further reflection. Within minutes I was involved in discussing the appropriate intervention for a patient who had reported obsessive thoughts about suicide. It was almost a pleasure for me to have to deal with this serious problem and the young therapist's fear that he had bungled the management of the declaration. At least I was dealing with suffering for which I was not directly responsible and had a real chance to reduce quickly.

So the reason I was spending as little time with the monkeys as I could was no mystery. The callus on my emotional perceptions was being shaved away, and my sensitivity to vulnerability and torment was becoming intensified. My clinical work, my ethics teaching, my participation on the psychology licensing board, my discussions with Bret and Charlene, and my work with the IACUC were all contributing to an increasingly honest assessment of the predicament the monkeys faced. That I was primarily responsible for this predicament presented a wrenching quandary.

Sitting in my office after the supervisory session had concluded, I found that the image of G-49 trying to squeeze a social life through a bolt hole would not go away. I retrieved a file containing letters written in response to a feature article about my work on the psychopathology of the social isolation syndrome in rhesus monkeys that had appeared in the *Albuquerque Journal* in 1982. I decided to reread the letters and let their contents surround me. I was not particularly interested in many of the letters, such as those calling the work a waste of money and effort. I focused my attention on the writers who expressed their concerns for the animals per se. These writers tried to put themselves in the monkeys' place and imagine what they were experiencing. In essence they described the suffering that they, the writers, were experiencing due to the knowledge that my experimental work and that of others was taking place and creating suffering in animals. The sentiments in these letters challenged the very reason I had gotten involved in research in the first place—to help alleviate human suffering. It was difficult now to face the fact that I was creating meaningful suffering in humans by doing my research.

After reading all the letters, however, I felt strangely comforted by their contents. The comfort arose from an awareness that I was going to have to make a decision about how to proceed in my research life. I knew I had to make a more affirmative response to my concerns and not just stay out of the animal lab and let graduate students do all the work. Some sort of resolution lay ahead.

A test of my changing values arrived some months later, when the dean authorized the Psychology Department to search for a new faculty member to fill a new senior distinguished professor position. Frank Logan, though no longer chair, was still a paternal force in the department, and his point of view carried a great deal of weight. Therefore when he declared a personal friend and professor from the University of Iowa, Isidore Gormezano, to be at the top of the small field of acceptable candidates, it appeared to many in the department to be a done deal. Professor Gormezano studied basic associative or Pavlovian learning processes in animals. He studied the way in which stimuli could come to control simple reflexive responses and what underlying brain structures were involved. He worked with rabbits, and his methodology involved delivering air puffs to the rabbits' eyeballs, measuring the reaction of the nictitating membrane, teaching the rabbits to associate some other stimulus with the air puff, measuring the impulses from the brain structures thought to be involved in the learning process, and creating brain lesions to test whether the associative learning could be inhibited. In the language of the day, he was searching for the "engram" or memory trace related to the learning and where it was located in the maze of neuronal systems.

I must admit that by this time I had lost a great deal of respect for Frank. I was inclined to resist his judgment of who was the best candidate from the very beginning of the search. During group meetings and presentations, Dr. Gormezano appeared to be a careful and thorough researcher devoted to the search for the mechanisms of learning and memory. Yet two things bothered me as I sat and listened to him. First, he seemed not all concerned by the fact that although human learning was certainly his overall interest, humans had only a vestige of a nictitating membrane, a fact that raised questions about the relevance of his work for understanding human associative learning. Second, he referred to the many rabbits that he used as "test tubes."

However, the most disturbing information came during a social gathering where he recounted the hassles of the airport delay that he experienced on his way to New Mexico. As I recall that story today, he told how a group of the stranded passengers sat together in a restaurant and began to exchange life vignettes. When it was his turn, he described his studies and his use of rabbits and brain-lesioning techniques. He was surprised when a number of the passengers looked horrified and dramatically got up and left the group. It seemed to me that he was absolutely perplexed by the negative reaction, even hurt by it. I could certainly understand that he believed his animal use was justified and important, but how could he be perplexed by some level of negative reaction? This was more than twenty years after the passing of the Animal Welfare Act and after the IACUC oversight system had been created. Was he as oblivious to the controversy and ethical debate underlying these changes as he appeared?

When it came time to vote on making an offer, I wanted to express my feelings and to do it quickly, before my need for safety overwhelmed my passion. By this time I had been promoted to full professor, and as a member of the senior faculty I knew that my word would carry some weight. However, I was more concerned with expressing my view of Dr. Gormezano and what he represented than with having an influence on the outcome. When the floor opened for discussion, I quickly raised my hand and was recognized to speak. I first expressed my reluctance to take a position that Frank might see as disrespectful of his judgment, because of my deep respect for him and his contributions to the department and psychology in general. This was a lie. I did not respect him, and I no longer valued his work. Nonetheless, I felt that this was a required preamble to my forthcoming attack. I went on to lay out my concerns about Professor Gormezano. I said that while he was no doubt an eminent scientist in his area, the title of distinguished professor called for a scholar with a broader set of concerns about psychology and society than I was aware that he had. While he was certainly a productive scientist, far more productive than I would ever be, I was surprised by his lack of first-author books and articles that attempted to present large integrated reviews. I then got to the heart of what I wanted to say.

"Society is becoming less tolerant of animal research, especially those experiments lacking direct or at least clear biomedical implications," I said. "Professor Gormezano's research raises all the red flags: survival surgery, unrelieved pain, restraint, references to rabbits as test tubes." I went on to characterize rabbits as organisms capable of experiencing various affective states like discomfort, satisfaction, and pain and to express concern about Dr. Gormezano's desire to "whittle away species and individual uniqueness and convert animals into basic 'preparations.'" As I expressed my position, the intensity I felt surprised me. It was one of the first times I had been honest with myself and my department colleagues.

Frank looked at me from across the meeting room with total disgust. However, neither he nor anyone else in the room said a word in response, either in support or in opposition. Instead, Frank just proceeded with his own presentation and spoke glowingly of the number of graduate students and the amount of grant funds that would come with Professor Gormezano if he took the position and moved to New Mexico.

In the end, Professor Gormezano was offered the job, but as I understood it, the salary negotiations failed and he stayed at Iowa. Frank never forgave me for what he saw as public questioning of his integrity. I was, however, thankful to him for helping to create a crossroads where I needed to decide whether to be quiet or speak out. We would hardly interact again from that time until his unexpected death in November 2004. I have no regrets.

*   *   *

Some time after the passage of the Animal Welfare Act amendments, I realized that the Psychology Department was in serious need of a graduate-level course on research ethics. My work with the Board of Psychologist Examiners had brought home the importance of issues of professional conduct, and exploring with students the ethics of research on humans and animals felt like just the right to be doing. Although the graduate curriculum already included a required course for clinical students on professional conduct, there was no course on research ethics per se for either clinical or experimental students. After studying in the area, I developed a course and offered it as a "special topic" elective because the department refused to add "yet another" required course to an already packed student schedule.

The course began with students reading and discussing articles on the history of researcher misconduct, mostly in the area of human research. We considered first the abuses disclosed after World War II at the Nuremberg Trials, discussing research showing that a host of academic scientists may have taken advantage of what might be called the largesse of the Holocaust by using tissues taken from executed prisoners and euthanized patients.

Next we discussed the important article by the Harvard anesthesiologist Henry Beecher published in the *New England Journal of Medicine* in 1966.[5] Beecher presented twenty-two examples of what he considered to be patently unethical human experiments. All posed great risk to the subjects without a shred of an informed consent process. The point of his article was not that there were always a few "bad apples" in a profession that needed to be identified and removed, but rather that the competitive environment of academic research was so intense that it inadvertently encouraged young researchers to cut ethical corners in order to accumulate the body of published data necessary to establish the basis for new treatments and professional advancement.

During discussions of Beecher's article graduate students often identified with his description of pressure and spoke of their own corner-cutting behavior and temptations to make their data conform to outcomes that would be publishable. They talked about rushing informed-consent procedures, not stopping to ensure that subjects truly understood the nature of the experiment in which they had agreed to participate, and underemphasizing (i.e., withholding) information that they thought might dissuade a subject from participating. Some even made reference to one of my senior faculty colleagues who refused to accept data that did not conform to his pet theories and required that they run the experiments until the data came out "right." Most students felt that the competitiveness of the current time exceeded what Beecher described for the 1960s, as fewer positions were available and expectations for publishing had increased significantly.

I invited Professor Archie Bahm, a senior faculty member of the Philosophy Department, to lead a discussion on the structure and application of ethical theory in everyday and professional life. The students enjoyed debating their values and conceptions of what constitutes right and wrong behavior. Professor Bahm communicated the position that ethics is a rigorous discipline and not just an airing of positional opinions, which appeared to be a common initial viewpoint among the students. For his part, he seemed to be genuinely interested in the ethical conflicts and dilemmas offered by the research environment, a topic that he had not before considered in depth.

These discussions led directly to an evaluation of the common practices involved in conducting human research. We studied and talked about the history of human research regulation in the United States that followed on the heels of the revelations about the Tuskegee syphilis studies in 1972 and other questionable research practices. We evaluated the Belmont Report, published in 1978 by the National Commission for the Protection of Subjects of Biomedical and Behavioral Research, which explicitly presented the ethical foundation of human research in the United States. The principles that the commission identified as relevant to the moral landscape of human research were the need for knowledgeable and uncoerced participation (respect for persons); prevention of harm (beneficence); and a just distribution of risks and benefits or research throughout the population (justice). In the last part of the course we read the "animal rights" philosophers, Tom Regan and Peter Singer, and their antagonists, such as Carl Cohen. Many students were surprised by the coherent arguments provided by the protection philosophers.

By many measures the course was very successful. I had no difficulty meeting required enrollment numbers, and student evaluations put it in the excellent range. I ended up offering the course nearly every year for a period of more than twenty years and often had to turn students away. After a couple of years the course began to attract students from other research-oriented departments like anthropology, economics, and exercise physiology. I was often stunned by the students' candor and their desire to discuss their ethical struggles—and to maintain their moral identities as a researchers, which they perceived as under attack by the competitive demands of being "productive." I felt certain that the course experience provided a safe and open environment for this self-exploration and was a valuable exercise for most students.

However, in another sense the course was a failure. During all the years I offered it, never once did an advanced undergraduate or graduate student who was primarily involved in animal research enroll in the course, despite the fact that there were many of these students in my department. I also did not attract animal researchers from the Department of Biology or of Anthropology, even though I sent information about the course to the department chairs and relevant faculty and often contacted them personally to discuss the plan of the course. While from time to time such students would arrive during my office hours to engage in spirited but time-limited discussions, none ever committed themselves to the in-depth study provided by the full course.

In trying to understand this phenomenon, I eventually came to the conclusion that human subject researchers were less likely to be overtly threatened by an ethical analysis of their practices. While they were certainly confronted by contentious ethical questions that might require that they alter their research plans and methods, the availability of human subjects for research was substantially grounded in autonomous persons' choice to participate. Even when there were adverse outcomes, the researcher could often claim that it was ultimately the subjects' independent decision to involve themselves. In addition, human researchers had the Belmont Report and its established ethical principles to help drive the preexperimental moral analysis.

Students involved in animal research, in contrast, knew at some level that once the ethics door was opened and the praiseworthy intentions to reduce human suffering and advance science were for the moment set aside, they would have to face the reality that animals were made available for experiments by acts of raw power. Animals were created, converted into products, purchased, shipped, and conveniently stored. Once owned, they could be modified and manipulated in ways generally unrelated to their interests as animals but according to our questions and current needs. This is dangerous territory for a morally serious person, so it is no wonder that animal researchers avoided going there in the first place.

*   *   *

By 1986 or 1987, it had become clear to me that I had to do something about the monkeys. The social group of stump-tails was not a problem; I could keep the group intact and limit the members' use to observation of their unique social behavior. The presence of rhesus monkeys in our facility, however, required justification, and I was unwilling to have them used for invasive experiments or to raise more of their offspring in social isolation. Moreover, my growing awareness of the poverty of their lives was proving more and more difficult to bear.

A partial solution to the problem arrived in the form of an inquiry from Mark Lewis, a neuroscientist from the University of North Carolina. He had become aware through Jim Sackett, at the University of Washington, that I was in the possession of a group of monkeys whose behavioral profiles included many stereotyped behaviors. Mark was a serious scientist who was interested in understanding the neurobiology of stereotyped behavior, which manifests in many human developmental disorders and institutionalized patients and is also present in monkeys raised in social isolation. Believing that research on my monkeys might help to unravel the causes and lead to useful treatments, he proposed that we enter into a collaboration that would have an excellent chance of receiving significant long-term funding from the National Institutes of Health (NIH). He had a particular kind of study in mind: after conducting more experiments on the isolate and control monkeys in my colony, we would sacrifice them and remove their brains for study.

As harsh as the proposal sounded, it would fulfill several contrasting wishes. First, the animals would eventually be killed, and I would not have to continually face the reality of their trapped lives and they would not have to live them. Second, the secrets of how the monkey brain was modified by social restriction would likely be revealed, resulting in a set of published papers and conference presentations. I was sure that these findings would underline the devastating effects of these types of experiments and provide further reasons why such experiments should not be repeated. I also knew that these publications would win the approval of my Wisconsin colleagues, as long as my real feelings about the ethics of the research remained mostly out of view.

After many conversations with Mark over a short period, I came away impressed by his competence, research plan, and clear concern that the monkeys not be unnecessarily harmed during the experiments. During the conversations I decided to tell him about my feelings that the monkeys were living difficult and ugly lives and that terminating them and their suffering had become a goal. He said he respectfully acknowledged my concerns. After considering all the various pros and cons, I agreed to the proposed collaboration.

If we received the NIH grant, it would provide me with a 30 percent salary raise in the form of a summer salary and other benefits, and it would contribute financial support for my graduate student Alan Beauchamp, a secretary, and a number of undergraduate assistants. These were all strong positives, but I was very aware of the hypocrisy involved. At the same time that I was beginning to express my concerns about animal research, I would be participating in and benefiting from the very research that I had begun to abhor. I rationalized that continued involvement in animal research provided me with a degree of cover about my real feelings. I knew that if the strength of my objections were widely known, I would be marginalized and would not be able to continue my involvement with the IACUC and related activities. In other words, the research gave me scientific standing and allowed me to try to improve the review system from within. While this is certainly an accurate description of my motivation, the cover also allowed me to hide from the responsibilities of my developing ethical position.

The brain study with Mark Lewis would allow me to provide twenty monkeys with painless and relatively dignified deaths. Many other rhesus monkeys lived in the colony, however, and they were actively producing babies. What should I do with them? I briefly examined the possibility of sending them to a sanctuary for former research animals. I inquired about several of these facilities and was left with the feeling that their financial stability was uncertain. I next called my graduate-school friend Stephen Suomi for advice. He was by then the chief of the Laboratory of Comparative Ethology at the National Institutes Health, which was not subject to the same financial uncertainties of independent academic laboratories and privately funded sanctuaries. We had continued to have occasional contact over the years at International Primatological Society meetings and at the dedication of the Harry Harlow Primate Laboratory. I thought of Steve as part of my Wisconsin community of friends and still felt the special bond that the Harlow laboratory had fostered among us. I had never directly discussed with Steve the ethical questions about animal research that had been stirring in my life.

After I described to Steve what I termed my "overpopulation problem," he offered to take the monkeys. Without any prompting from me, he added that he would make sure that they stayed in their social group and would not get "chopped up." I understood this assurance to mean that they would not become subjects in neurophysiological experiments. I felt incredible relief. I would need to send their medical records to his veterinary staff for final clearance before the transfer could happen.

Once all the paperwork was completed, I contacted a professional animal transporter named Earl Tatum and inquired about how best to move the animals from New Mexico to the Washington, DC, area. I dreaded the thought of sending the monkeys individually as air freight. Tatum, who shipped many exotic animals for zoos, told me that would not be necessary. He said that he had a truck that would hold all the monkeys, and he would drive nonstop until they were delivered. Since he came highly recommended by local zoo personnel, we made an agreement to proceed.

When Earl and one other man arrived, they wasted no time and moved directly to my outdoor facility. They walked slowly around the perimeter, eyeing the animals like members of a street gang sizing up their opponents. Earl asked me to identify the dominant male and female, and the answers verified his own conclusions. The two men then put on wrist-high leather gloves and entered the facility with a large net and two transport cages. They immediately trapped the dominant male and female and removed them from the area. Once this part of their plan was complete, they proceeded to herd the remaining individuals into transport cages. The move proceeded with such speed that there was remarkably little resistance and screeching and no serious injuries. The animals were then moved to his truck, where they were released into an open area about one-third the size of the outdoor facility from which they had just been removed. It contained a full complement of perches, floor space covered in hay, water containers, and food bins. The monkeys quickly located themselves throughout the space and sorted themselves into their typical clusters. I signed the release and they were off. Unlike the surgeries yet to come, it all happened so quickly that there was hardly a moment to look at each monkey separately and whisper a goodbye. I watched as the truck left the parking lot of the psychology building and made a right turn in the direction of Interstate 40, which would take them all the way to the East Coast. Two days later I was notified that the animals had arrived at the NIH facility safely.

Part of my wish to be free of the monkeys and my responsibilities to them had been achieved. In that moment it seemed good: Steve Suomi would not run out of money to care for them, and they would live out their lives in a social group being noninvasively observed by researchers interested in primate social behavior.

I went up to the roof and sat by the empty enclosure. With the monkeys gone, I could now afford to think about the individuals that I had gotten to know over the years. E-3, the dominant male with the fuzzy face, was the gentlest rhesus monkey group leader that I had ever seen. He huddled with the young ones and groomed his favorite female, Alice, constantly. With him as leader, the group had become relaxed, with very little fighting. There were so many stories and moments of discovery about the complexity of their social and emotional lives that I had access to. I wondered whether I had made a humane decision for both the monkeys and me. Had I sold them down the river just so that I could feel less pressure and begin to create a different identity?

By 1988 we had concluded all the NIH-funded monkey experiments, and now it was time to face the final phase of the study. The differentially reared monkeys would be sacrificed and their brains studied. Mark had invited Dr. Linda Cork, a veterinary neuropathologist from Johns Hopkins University, to direct this portion of the grant, and I had invited Steve Shelton at the University of Wisconsin Medical School to do the studies on the content of the cerebral-spinal fluid that would be collected at sacrifice. We decided on a date, and all the travel arrangements were made. Now I knew this would really happen. As if in a long, mind-numbing military march, I went through the steps of scheduling the appropriate lab space in the medical school and acquiring all the necessary drugs and chemicals required. Luckily my friend Phil Day, the campus veterinarian, knew what exactly was required to anesthetize and kill twenty monkeys. He arranged for all the equipment: scalpels, surgical cutting saws, gloves, masks, gowns, dry ice, and shipping crates. He also modeled the necessary attitude for the event: nonconversational proficiency. Now care for the animals needed to be expressed by doing everything right so as to ensure that they would not suffer pain at the end and that they would provide good data about the anatomy and chemical architecture of their brains. I decided that although I had no skills relevant to the removal of brains and prepping and packing them for shipment, and all the anesthesia inductions would be done by the team of Mark, Linda, and Phil, I still wanted to be present for each procedure. So I specified that I would give the final OK to proceed once I was satisfied that each animal was deeply anesthetized. The process would take five days, with four monkeys being sacrificed each day.

On the morning before the first monkeys were scheduled to be sacrificed, each received a special portion of his or her favorite fruit several hours before being weighed, placed in a transport cage, and driven to the Medical School. There Ector Estrada, the technician who was most familiar with each animal, restrained them with his firm, gentle method while they were administered a preanesthetic restraint drug. Once they were able to be easily handled, either Ector or I carried them to the surgery and placed them on the operating table under a chandelier of bright surgical lights. Phil then administered the anesthetic drug while the other members of the team prepared themselves to proceed. Linda checked the status of each animal by looking for a response to a hard pinch to the skin between the toes, and for a blink reflex when she placed her index finger on one set of eyelashes. Once she was satisfied that the monkey was in a deep anesthesia, she nodded to me. I then did additional tests, and if they proved consistent, I simply said "Go."

As the procedures went on, I could tell that the team was becoming impatient with my reliability checks and the delay that they caused. We were about to do E-25, a female partial isolate, who had been born at the Wisconsin Primate Lab twenty-two years earlier. She was among the first animals that Harlow had shipped to me for my rudimentary lab in the hallway of the Chemistry Department basement in 1971. She had participated in many of my experiments, and I thought fondly about her gentle and cooperative manner. Locked in these thoughts, I was being particularly slow in giving the go-ahead. I heard Linda sigh deeply and tell me in a sing-songy voice, "She's ready." I retorted that I thought I saw a weak blink reflex. Again the sigh from Linda. Next I opened E-25's mouth and inserted my index finger deep into her throat. She gagged. She was *not* completely unconscious.

After this incident, the induction process became less hurried and complaints about my methods stopped. Cutting through the skull of an immobile but partially aware monkey was the last thing anyone wanted to see happen.

Once there was agreement that a monkey was unconscious, Linda used an oscillating bone saw to cut horizontally through and around the skull, beginning at the level of the eyebrows. The brain was then "shucked," prepared, placed in ice, and made ready for shipment. While the team focused on the monkey's head and the removal of its brain, I snipped a small clump of hair from its thigh and placed it into a letter-size white envelope in my lab coat. I tried to do this surreptitiously. I didn't know what I was going to do with the collection. I had only a partially formed thought about keeping a remembrance or conducting some kind of tribute.

I had read a good deal about the rituals of apology that were part of many different aboriginal hunting practices.[6] The idea, as I understood it, was that if the animal was properly respected, it would offer up its life willingly to the hunter, and its spirit would remain and enter the body of another animal and then return to be taken again. If treated with disrespect, the animal would not return. I had once witnessed a man from the Isleta Pueblo, just south of Albuquerque, perform an apology ritual after killing a deer. He knelt next to the deer and thanked it out loud for the gift of its body. He then cut a small wedge of flesh from the thigh and tossed it into the grass, "returning the swiftness and spirit of the animal back to nature." After gutting the deer, he sliced off a small piece of its liver and gave each of us a piece to eat, "joining our bodies together as brothers." While I didn't believe that the deer suffered less than the animals I saw my rancher friend Charley Oates kill, I did understand the apology ritual as an attempt to acknowledge and reduce the tension produced by the existential reality of life needing to live on other life. The custom gave pause to the death and emphasized that while something was gained, something important had also been lost. I wanted to express sorrow for taking the lives of the monkeys—not for terminating their lives, but for using them for my own purposes and thereby negating their inborn and developmental promise.

At the end of the week-long process, the dominant feeling was relief. While I tried to stay present to the emotional reality of what was happening animal by animal, a certain mechanical rhythm developed that on its own created a distance. I found that it required intense mental effort to stay totally involved. At some level I wanted just to walk away and return when it was all complete. I had to nail my feet to the floor and remind myself over and over that guarding against a last-minute harm was my last opportunity to do something positive for these monkeys. While my insistence on rechecking evidence of consciousness had perhaps prevented several animals from experiencing some moments of excruciating pain, there was no ritual of apology that could set right what had been done.

When the results of the terminal studies were finally known and published in a series of articles beginning in 1989, they emphasized just how damaging raising monkeys in socially deprived laboratory environments is from a biological perspective. The evidence was clear that the brains of the socially deprived monkeys were structurally and chemically different from the brains of the socially reared monkeys, and the cell-mediated arms of their immune systems had been changed. Analysis of the survival rates of the entire group of monkeys that I started with in 1971, apart from the ones that had been euthanized for the neurological studies, revealed that isolate males died first, followed by the females, then the socialized controls. There was also suggestive evidence that the isolate monkeys suffered from more frequent respiratory and intestinal infections. The changes in the immune system also suggested that the isolates may have suffered from many more subclinical illnesses than the socialized controls. In other words, the effects of isolation ran very deep, beyond just obvious behavioral alterations. These findings pointed to a laboratory life that was even more uncomfortable than I had already come to appreciate. Adding the costs of frequent illnesses was very much unexpected.[7]

*   *   *

In the summer of 1990, while I was still engaged in helping to analyze data from the brain study and write up the results, Mark Rutledge, the University of New Mexico campus minister, invited me to give a presentation in his popular open university Last Lecture series. The idea of the series was to give faculty members an opportunity to convey to a general audience of students, faculty, and members of the public the one message they would want to leave behind for the world. I took the "last lecture" concept to mean that I was to imagine my impending death and allow that perspective to cut through the cover of the "just interesting" to the heart of my beliefs. There were no other constraints. I accepted the invitation without knowing what I would lecture about, recognizing it as a wonderful opportunity. The rush of academic achievement rarely encouraged such reflection.

As the date approached, I considered many topics. Among my possible titles were "The Keys to Effective Personal Change," "The Source of Human Potential," "The Recovery of Professional Conduct," and even "The Mind of a Monkey." I was connected to these topics in serious ways but felt that they didn't quite suit the occasion. I doubted that they would be among my deathbed messages. As the date approached and the need to advertise arrived, Mark pushed for a decision.

I was having a discussion with him on the phone about the options when I was interrupted by the arrival of a biology graduate student with whom I had a scheduled appointment. The purpose of the meeting was to transmit a number of questions from the university's IACUC about a research proposal that the student had submitted. The proposal involved trapping a particular species of bird during its breeding season with the use of mist nets, drawing blood samples and taking various physical measurements, and then hopefully releasing them unharmed.

One set of questions on the list involved what plans the researcher had in case a bird was seriously injured during netting and handling. The young man indicated matter-of-factly that he would quickly euthanize such individuals by either breaking their necks or squeezing them tightly in his hand to suffocate them. I reacted with doubt to the suffocation procedure, knowing that there was evidence that the death it caused was prolonged.

I next asked about the likelihood that the captured birds would be tending nests of eggs or newborns and what would happen to them if the adults become disoriented from the netting and handling or had to be euthanized because of injury. He stared at me with a pained look and repeated the question but with a slight change. "You are asking what will happen to the nest contents?"

"Yes," I responded, "what will happen to the eggs or baby birds?"

"The nest contents?" he repeated in a stronger tone. "They will disappear."

"Where will they go?" I asked. "Won't they just starve to death?"

"I guess so," he conceded.

I asked why the project couldn't be done outside of the breeding season, when such unintended tragedies would not occur.

"Because that is when I have the time to do it." After making the statement, he quickly amended it by saying that he was interested in the birds' breeding condition.

This was not unlike many of the unsatisfying interactions that I had as the chair of the IACUC; the committee would try to push for even a slight minimization of animal pain and would encounter frequent resistance based in methodological tradition.

The conversation was brief but poignant to me. Here was this smart and apparently decent student resisting expressing directly the costs of his research in terms of animal harm. He preferred to cleanse his descriptions and minimize the reality that real animals, living a full but no doubt difficult natural life, would experience struggle, real pain, and possibly death in the service of his questions about nature.

After this conversation and a little quiet time, I called Mark back and said that I was ready to commit myself to a topic. It was time to take thinking about animal issues beyond a small classroom and a tirade against a possible faculty hire in a closed departmental meeting. I offered up the title "Why Is My Brother in a Cage?" Mark seemed surprised by my choice but not disappointed. He suggested that instead of "brother" I use a nongendered descriptor that still expressed the idea of kinship. In the end we settled on the title "Why Is My Cousin in a Cage? Arrogance and Responsibility in Animal Research."

Planning the presentation was surprisingly easy. I would discuss Western culture's domination of nature and animals, the Judeo-Christian idea of improving the world, the desire of scientists to reduce the pain, fear, and suffering of other humans, the war metaphor used to frame medical research, the ways in which serious moral consideration of the costs of animal research is deflected, the institutional momentum of the animal research enterprise, and the tendency of researchers to see the animal protection movement as composed of only anti-intellectual emotionalists. I would attempt to leave the audience with the idea that because we don't really know what animals are capable of cognitively and in terms of pain and suffering, we cannot be certain of what the costs really are when we use animals in research. My overall message would be that if a researcher does not at least feel strongly ambivalent about using up the lives of animals to assuage our own suffering and fear of death, an ethical tragedy has taken place. My plan was to conclude with a quote from the philosopher Hans Jonas: "Let us not forget that progress is an optional goal, not an unconditional commitment. . . . A slower progress in the conquest of disease would not threaten society . . . but society would indeed be threatened by the erosion of those moral values whose loss . . . would make science's most dazzling triumphs not worth having."[8]

When I stepped up to the lectern and looked out at the audience, I did not see anyone I knew save for the daughter of one of my departmental colleagues. Once I got started, my presentation style felt different from the usual. There was more of a rhythm as I moved through the sections, and the volume of my voice glided from soft to assertive in a way that was unplanned and unfamiliar. At one point during the presentation I talked about H-89, a female stump-tailed monkey who, because of her intelligence, had become my partner in evaluating my learning tests before I used them formally in an experiment. Over time I had found that if there was a way to beat the system and get rewards without solving the problem legitimately, she would find it. My point was that as much as I thought I knew about monkey learning ability and motivation, I did not really understand who she was. She seemed to be an intentional being who derived pleasure from making a fool out of me. If this was true, what ethical implications did it have? As I told the story during the lecture, my eyes began to tear up and my voice became thick. A chill spread over my torso. The emotion arose because in that moment I was experiencing the importance of her existence independent of scientific use, and in part because H-89 had died a terrible death for which I was partially responsible.

As I approached the conclusion of the talk, I became aware that the room had become very quiet and that I felt exhausted. The typical sounds of restlessness and disinterest, always a part of even the best-planned lecture, were virtually nonexistent. Something unique had happened. I believe that my presentation had tapped into the reservoir of natural interest and concern for animals that rests under the consciousness of most morally serious people and had allowed it to flow around the everyday defenses that usually limit their awareness. For a few moments, the profundity of the moral questions about the use of animals for our benefit became vivid for many members of the audience.

Afterwards, the deep satisfaction I felt in the wake of the lecture was tempered by the realization that my intellectual passion and moral identity were increasingly diverging from much of the rest of my academic life.

* * *

In 1992 I was made acting chair of the Psychology Department. The additional administrative responsibilities helped justify to myself and others what at that point had become an almost complete withdrawal from lab animal research. On the other hand, the very public nature of the leadership role made it more difficult to hide from the critical spotlight that some faculty were beginning to train on me now that my views on animal research were more or less out in the open. Fortunately, the publications that came out of the brain studies made it difficult for people to clearly identify me as a budding animal protectionist.

It was around this time that I was forced to confront an uncomfortable consequence of the brain study work. As more data from the studies of the brains of the isolates and controls began to appear, I received a question from the neuropathologist, Dr. Cork, about whether I had any behavioral data on the animals who had been sacrificed. I responded that I had data on learning performance and levels of stereotyped behaviors like rocking, huddling, and self-mutilation. She asked whether she could have them so that she could try to relate the brain changes she was seeing to these behavioral measures.

This sounded like a good idea. However, after we exchanged more questions, I learned that the overall plan was to use the data to apply for a grant in order to raise a new cohort of isolate monkeys that would be sacrificed at various time intervals during their isolation confinement and afterward in order to determine at what point the brain changes seen in our adults began to occur developmentally. Dr. Cork mentioned that Jim Sackett was likely to be a coresearcher on the project. I was absolutely stunned by these plans.

From the beginning of the project, I had partially rationalized my participation on the assumption that studies involving the social isolation of monkeys had been recognized as now being beyond the ethical pale, but that salvaging information from my animals was a way to extract some good from the suffering they had endured. I had expressed this perspective to Mark Lewis as we were initially exploring the possibility of submitting the grant together. I felt that he at least respected my view. So here I was finding out that the work was instead going to be used to scientifically justify yet another round of separating infants from their mothers and raising them in sound-proof boxes for many months. In trying to justify my participation and accumulate the benefits of being a federally funded investigator, I had allowed the pretense that the members of the research team all agreed with my position to stand. Social psychologists describe this phenomenon as the "false consensus effect"; I call it lying to oneself out of self-interest.

What was I going to do? From a purely scientific perspective it seemed that I could not deny them the use of the data. From an ethical perspective, however, I had reached my limit. I could not directly participate, any more than I already had, in facilitating new isolation studies. But didn't I owe something to Jim Sackett? Jim had made graduate school possible and had taught me the thinking and methodological skills required for scientific competency. Would he see my resistance as turning my back on him? Would he see all the effort that he invested in me as a waste? In other words, was I going to lose a friend and a major source of my professional identity?

I briefly considered a deception where I would blame the US Postal Service for failing to deliver the package of records, and a secretary for not copying the data sheets before mailing. I rejected the option as not believable and not worthy of my ethical struggle. I was on my own in a new way. Finally, I wrote to Dr. Cork and told her that while I understood her plan and respected Jim as a long-term colleague and mentor, I would not send the requested information and thereby support the design of more isolation studies with monkeys.

I don't now recall hearing any rejoinder to my refusal directly from Jim. However, somewhat later, Alan Beauchamp—by then a new assistant professor at Northern Michigan University—informed me that Dr. Cork had asked him for *his* data to achieve her purpose. I immediately wrote an angry letter to her objecting to what I believed was an end run around my commitment to not contribute to more monkey isolation studies. She denied such a purpose, but I was not convinced. I reminded her that those experiments had been done in my laboratory and I was a coauthor on the publications.

I was coming to realize what it was going to require to transform my disturbing feelings about monkey isolation research into a visible stand. Changing my research focus quietly was no longer an option. Alan and I stood by our decision not to provide the requested information, and surprisingly I did not hear another word from Dr. Cork.

Two other incidents occurred while I was acting chair that had the effect of keeping the animal issues on the front burner. During the first week of my term, a detective from the campus police confronted me with his worry that some controlled drugs held in the animal facility might be being misused. Stating that there had been at least two reports in recent years of drug thefts from those facilities, he ordered me to audit all departmental purchase orders for controlled substances and then to see whether the items purchased were requirements of an approved research protocol.

The search revealed purchases of pain medications that were not part of any animal research protocol. The person identified by the audit was then confronted in my office by officers from the federal Drug Enforcement Agency (DEA), the Board of Pharmacy, and the campus police. The person immediately admitted to the infractions, including seriously deficient recordkeeping, and surrendered the government license he had used to purchase the controlled drugs. Later I learned that rumors had circulated immediately thereafter that I had "set up" this person because of my dislike of his pro-animal-research views.

Shortly after this incident, the campus veterinarian brought a complaint to the IACUC about the death of a number of mice and rats, all belonging to the same researcher. Pathology reports showed that the animals did not show signs of infectious disease but had no discernible fat stores anywhere in their bodies, indicating that they had starved to death. The vet reported that while investigating the deaths of the rats, he found that many other rats in the same lab were alive but emaciated. Their spines protruded, their coats were rough, yellow-tinged, and ungroomed, and they held themselves in a contorted, hunched posture. These characteristics screamed suffering as clearly as if the animal could speak English. He said that he was surprised that they, too, were not dead.

Unfortunately, instead of expressing surprise and concern, the researcher immediately became very defensive and insisted that nothing was done inappropriately and that he was using acceptable experimental methods. An examination of the paperwork and the experimental design revealed that the researcher had failed to include his plan to reduce the animals to 80 percent of their free-feeding weight in the formal research protocol previously submitted to and approved by the IACUC. Not only that, but the researcher had determined the baseline weight from which the rats were to be reduced early in their lives, before they had reached their full adult size. As a consequence, the animals had actually been reduced to between 50 and 70 percent of their likely adult free-feeding weight. The veterinarian noted that the fact that more animals had not died was a tribute to the resilience of the laboratory rat and the adaptations that permitted it to survive in the face of such desperate nutritional conditions.

How could a senior researcher not comprehend these weight relationships and their effect on the animals' health status? Wasn't it likely that the emaciated condition of the animals negatively affected the validity of the experimental results? Was this the result of carelessness, a poorly supervised staff, the intent of a sadistic rogue technician? The investigator continued to deny that anything had been done incorrectly.

As chair of the IACUC with jurisdiction in the matter, I recommended the suspension of the research pending a full investigation. While amazingly there was some sentiment on the committee that the situation did not constitute a serious problem, the majority of the committee did eventually vote to suspend. This action became widely known across campus and surely must have provided more evidence to some that I was out to discredit those involved in animal research. The problem disappeared into the upper administration and was joined with other complaints about inappropriate sexual relationships with students. The laboratory closed, and the researcher left the university. Later, an e-mail was circulated to those of us with knowledge of the issues that directed us to forward any questions about this person to the upper administration while giving a "no comment" response to the inquirer.

During this time I began to read in more depth about ethics and animals. I found the work of Professor Bernard Rollin at Colorado State University[9] particularly useful. Then I discovered an edifying article on the moral standing of animals by the philosopher Tom L. Beauchamp.[10] Beauchamp quoted an interchange between the philosopher Robert Nozick and the Nobel-winning biologist David Baltimore that was a response to the film *Primate* by the documentary filmmaker Frederick Wiseman. Wiseman had filmed what was supposed to be typical research activities taking place at the Yerkes Primate Center in Atlanta. The film depicted the living conditions and work environment of the great apes and monkeys held in that facility. I had not seen the film, so I purchased a copy and viewed it. I thought that it quite accurately captured the work flow and appearance of a 1970s primate laboratory. The caging was clean yet barren, the researchers serious, and the costs borne by the animals in the service of research obvious. This was a different kind of exposé from one secretly filmed. With much of the intellectual context of the work stripped out, the researchers seemed a bit foolish and the animals hopelessly vulnerable.

In the exchange recounted in the Beauchamp article, Baltimore simply said that he did not see anything in the film that concerned him, while Nozick expressed shock that the pain and distress obviously being experienced by some of the animals seemed not to be a factor in the justification of the research. Like Wiseman in the film, Beauchamp allowed the principals to speak for themselves and let the conflict frame itself.

I also found several papers on the ethical obligations of the IACUC by F. Barbara Orlans that addressed some of my own concerns about the system.[11] Both Beauchamp and Orlans were faculty at the Kennedy Institute of Ethics at Georgetown University. This fact planted Georgetown firmly in my consciousness. Then, several months after I had given my "Why Is My Cousin in a Cage?" presentation, I came across a comment in the journal *American Psychologist* by Daniel Robinson, chair of the Department of Psychology at Georgetown, about how his faculty had decided that the typical defenses of animal research in psychology were mostly "self-serving and entirely unconvincing" and had therefore halted the research and closed the animal facilities.[12] I discovered that the Kennedy Institute of Ethics had a visiting fellows program that provided opportunity for concentrated study with a faculty mentor along with a general exposure to the rapidly expanding field of bioethics. Once I read about the program, I knew it was exactly what I wanted to do.

Although I was serving as the acting chair of the department, I was eligible for a sabbatical leave beginning in the late fall of 1993. After I talked to Charlene about the opportunity, she agreed that I needed to go to Georgetown. I applied for the fellows program, requesting to work on animal research ethics with Tom Beauchamp and Barbara Orlans, and received a quick acceptance. I then filled out the sabbatical leave application, indicating my plans to study bioethics with a focus on animal research, and sent it along to the college committee responsible for approval. I timed the submission perfectly and received a rapid response—disapproved!

The committee, made up of senior faculty from the College of Arts and Sciences, had apparently decided that they could not see how an educational experience in bioethics and research ethics would possibly benefit either me or the university at large. What I needed to do, they suggested, was to seek an experience that would impact my empirical experimental work and thus my ability to garner more extramural research funding.

I had allowed myself to assume that the profundity of the ethical questions related to animal research were just obvious. I believed that the hostile attitude I had seen many years earlier, when I dared to bring up the issue of the possibility of research misconduct by Sir Cyril Burt in my debate with Professor Gaynor Wild, had since dissipated as researchers and scholars became enlightened. I was wrong.

I made an immediate appointment with Dean William Gordon. During our meeting I expressed my disappointment with the committee's decision and reiterated my justification. I also cited recent examples, of which he was also aware, of sabbatical leaves being granted to faculty with vague plans like "developing a book" or "analyzing data that had laid fallow." I explained that upon my return I planned to create a research ethics consultation service and showed him evidence of similar services being developed at elite universities around the world. He was clearly sympathetic. As a final touch, I told him that if the sabbatical was not approved I intended to file a complaint with the Academic Freedom and Tenure Committee and go immediately on the job market.

This threat violated a major piece of advice that Harry Harlow had given me just after I graduated from Wisconsin and before I left for my first job. "Never threaten to leave an academic job if you are not imminently ready to pick up and go. If your bluff fails, you have lost all your leverage with the administration from then on." I was not in fact ready to leave, but I was sufficiently incensed about the denial to take the risk.

In what seemed to me a bargaining move, the dean asked if I

would be willing to entertain appointment as the regular chair of the Psychology Department once I returned from the leave. I agreed that I would, and the sabbatical was approved shortly thereafter. Not only was it approved, but the dean generously provided additional financial support to ease the cost of relocating for a year. I was eager to begin this next important step in the process of altering my academic and professional identity.

6

Reconstruction

One may ask even of the devotee of science that he should acquire an ethical understanding of himself before he devotes himself to scholarship, and that he should continue to understand himself ethically while immersed in his labors.

SØREN KIERKEGAARD

When I walked on to the Georgetown University campus in Washington, DC, for the first time in early January 1994 to begin the one-year fellowship in bioethics, I immediately encountered St. Mary's Hall, a plain red-brick building that housed the College of Nursing. Set into the north-facing wall was a larger-than-life sculpture of the Virgin Mary in white granite, her hands pressed together in prayer and her head and shoulders bowed slightly, draped with a long shawl. When I looked closely at a hazy brown area on the white stone, I could see that a pair of birds had built their nest in the space between her hands and her chest. That the nest was not taken to be a defacement and removed made me smile and think that I had come to the right place to study the ethics of animal use.

I proceeded through the campus, walking past the hospital, a classroom building, and a cemetery where deceased Jesuit brothers were buried. The cemetery's proximity to the hospital, classrooms, and research facilities set a realistic tone about the limits of medical healing and the ultimate course of life.

I entered the small, pitched-roof building named Poulton Hall and climbed the steps to the second floor, where the Kennedy Institute of Ethics was located. It was a simple place, not the grand edifice that I had imagined for an institute named for Joseph and Rose Kennedy. I found the conference room where the initial meeting was scheduled, sat down, and waited. Soon a trail of others began to arrive, and after a few moments we broke the awkwardness with energetic introductions. The class of thirteen visiting fellows was varied with respect to professional and international identity. There were four physicians; two doctoral-degree nurses; three philosophers, two of whom were Catholic priests; an anthropologist; an official of the World Health Organization; a journalist; and me. For a brief time, Dr. Bernard Nathanson, narrator of the anti-abortion documentary *The Silent Scream*, was with the group, but he did not stay beyond a month for a reason that was not disclosed to us. The countries represented included the United States, Argentina, Denmark, Germany, the Philippines, and Japan. As explained to us by the executive administrator, Irene McDonald, we were to meet formally as a group once a week for a three-hour seminar on ethical theory and bioethics, choose from a selection of ethics-related courses in the Philosophy Department, attend hospital ethics committee consultations, and meet with our mentors at least once a week.

During the initial group meeting we took turns describing our professional backgrounds and goals for the fellowship. It was fascinating to hear my new colleagues present their interests, which included the process of ethical decision making at the end of life, the ethics of the new genetic and reproductive technologies, the meaning of medical futility, the ethics of euthanasia, theories of justice, the welfare state and health care, physician paternalism, the ethics of care, advance directives, the definition of death, Japan's wartime human experiments, and health policy related to the treatment of indigent immigrant groups.

When my turn came, I characterized my interests as research ethics in general and the ethical justification of animal research in particular. I scanned the group for their reaction. Did they consider my goals to be as worthwhile and interesting as I found theirs, or would there just be some inauthentic reflexive head-nodding? Even with the sensitivity of my clinician's eye and ear turned up to high, I saw only honest affirmative reactions. There were spoken expressions of support and even some suggestions for philosophical readings. It was clear to me that my cohort saw the animal issue as important.

After I met with my mentors, Tom L. Beauchamp and F. Barbara Orlans, later that first week, I knew for certain that I was at the right place. Tom's confident presence and the pace and content of his measured speaking style suggested a serious, methodically thorough man who didn't waste time. He was welcoming and expressed strong interest in our working together. As I learned, Tom was a David Hume scholar and well known in bioethics for his work in the mid-1970s on the National Commission for the Protection of Human Subjects of Biomedical and Behavioral Research. In connection with the latter, he was the primary author of the Belmont Report, a seminal document in which he articulated a set of ethical principles (respect for persons, beneficence, justice) designed to guide ethical decision making in human research. Beyond this, he had written broadly on death and dying, informed consent, and the moral standing of animals. He authored the widely used textbook *The Principles of Biomedical Ethics* with James Childress from the University of Virginia. Barbara was a physiologist educated at the University of London who, after moving to the United States, took a job at the NIH, became interested in ethics, and then joined the institute. Her friendly smile and bright eyes were made even more charming by what remained of her British accent.

At the conclusion of the first meeting, it was decided that I would begin reading the recent book *The Great Ape Project* by Paola Calavieri and Peter Singer,[1] as well as works by several philosophers on the topic of personhood. With my assignments made, Barbara and I moved to the institute's conference room to continue our conversation. We opened up to one another immediately. When I asked how she had moved from being an animal researcher to a person devoted to animal welfare and associated ethical issues, she described a set of experiences that I could easily understand. She said that when she began working at NIH, she was in a lab where one of her colleagues, who was studying the physiology of cellular mitochondria, started each day by picking up a fully conscious mouse by the tail, dropping it into a Waring blender, and liquefying it. All the movements, from removing the mouse from its cage to dropping it in the blender, were done, she said, without the slightest hesitation or apparent thought for the animal's admittedly brief but intense pain. She also described being horrified by the housing conditions of the experimental dogs she encountered at the facility. Until then she had not realized the extent to which the standards and regulation of animal lab practices in the United States were inferior to what she was familiar with in the United Kingdom. Unlike the case in Britain, animal caretakers here were mostly untrained, researchers were not required to take in-depth courses on animal welfare methods and pain control, and there were no laws that provided for oversight of experimenter expertise and experimental goals.

Finding herself in the midst of this situation, she had felt compelled to directly voice her concerns and quickly became known as "the animal lady," an epithet not meant as an appreciative acknowledgment of her interest and kindness. She felt that many of her fellow scientists saw concern for the welfare of animals as close to silly, certainly unnecessary, and the proper interest of only nonprofessional women with too much time on their hands.

She then described the impact that attending science fair exhibits around the Washington and Maryland area had had on her animal-dependent science education concerns. These fairs included the prestigious Westinghouse Science Talent Search and the International Science and Engineering Fair. She had come upon entries by junior high and high school students that entailed doing brain surgery on monkeys in a garage, feeding toxins to pregnant rabbits and documenting the deformities of the offspring, spinning mice on turntables for days on end, and so on. She felt that these children were doing projects far beyond their level of knowledge and skill and as a consequence were being taught to be insensitive to animal pain and to see this insensitivity as a necessary part of being a research scientist.

Historically, she said, several important research institutions fostered this kind of early childhood involvement in animal science. The National Cancer Society once circulated literature about how schoolchildren could help "cure" cancer by doing home experiments on tumor-ridden mice provided by a local scientist, and NASA had encouraged students to send up mice in homemade rockets. As she summarized this history, her eyes flashed with passion and pain. Barbara was in no way antiscience, nor was she a Luddite fearing a technologically altered future. She was simply saying that the justification of research was a balancing act between potential advances and the costs in pain borne by the animal subjects, and that all too often this ratio was too unbalanced to proceed. Losing one's sensitivity to the predicament of the study animals left the process bereft of an ethical foundation.

As she talked about the science fair issues, I remembered the time I had served as a reviewer for a science fair held on the University of New Mexico campus in the 1970s. In one entry, a high school student had subjected a number of rats to burn trauma and then studied the physiological consequences. Even though at the time I had a relatively uncritical stance on animal research, I was surprised that a child was conducting such a sophisticated and pain-producing study. Others on the committee assured me that the study was in fact being conducted in the child's parent's university laboratory and was part of their grant-supported research. Apropos of Barbara's concern that many studies served primarily to desensitize the participating student to doing harm in the name of science, I remembered that the project's title made reference to the effects of "thermal stimuli" on rat physiology. Anyone would know that a burn hurts, but the phrase *thermal stimulus* made it more abstract to the reviewers and perhaps to the young student as well.

With neither of us wanting to end the conversation, I began to tell Barbara about myself. I recounted my comfortable sensitivity and concern for animals as a youth, and then my vivid desire to become a psychologist-scientist studying learning and animal models of psychopathology. I told her about the neurological and mental illness in my family and my wish to benefit others similarly afflicted. I described my early recognition that working with animals required a thick skin and my desire to gain this badge of rigorous objectivity even though it went against my nature. I explained to her how easy it was to ignore or compartmentalize the reality of animal pain and the harsh treatment I delivered to experimental animals, and how my desire to "find out" obscured and weakened the concern about harming. And how the changes that occurred in the process of desensitization were camouflaged by intense curiosity, outside encouragement, and ambition.

I talked about how the ambivalence had begun to creep back in, coming through the openings supplied by students' questions and the clinical training that made me aware of the importance of seeing the animals' lives from their point of view. I explained how I began to pursue a less ego-involved assessment of the value of my work. I recounted my disappointment at what I saw once the Animal Welfare Act and the IACUC system cracked open the laboratory doors starting in 1985. I admitted that initially what I could not see too clearly in myself I could see more easily in others. I told Barbara about my now strong desire to change direction, away from animal research and toward a competence in applied ethics. I expressed hope that my experience as an animal researcher might provide me with some legitimacy and added influence when I encountered other professionals.

Although I could feel that a bond was forming in those early hours of our relationship, I did not bring up the last monkey studies done with Mark Lewis and Linda Cork. I did not speak of what I now knew was the abandonment of the monkeys I had sent to Steve Suomi's NIH lab in Poolesville, Maryland, only miles from where we were talking. I wasn't sure she would understand. After all, I wasn't sure I understood it myself. With my eyes half-open, my heart mostly closed, but a growing interest in ethics stirring in my intellect, I had sent the monkeys under my care to a lab where I knew deep down they would ultimately be dismantled as a social group. I could hardly face that reality myself.

The year that followed was the most important of my post-PhD professional life. The weekly seminars, led by some of the most thoughtful and respected bioethicists in the world, provided exposure to a broad variety of bioethics issues. People like Robert Veatch, Ruth Faden, Leroy Walters, Edmund Pellegrino, Margaret Little, Dan

Sulmasy, Madison Powers, and Henry Richardson lectured on end-of-life issues, research misconduct, quality-of-life determination, health policy, distributive justice, and hospital and clinical ethics. In the classes on ethical theory I read Aristotle, Jeremy Bentham, J. S. Mill, John Rawls, Kant, Robert Nozick, Ronald Dworkin, Peter Singer, and many others.

My weekly meetings with Tom Beauchamp were extraordinary in their intensity and focus. We began our discussions with the central concept of moral status—that is, asking what characteristics an entity must possess in order for it to be considered an object of moral concern and thereby gain access to some level of protection under the ethical norms of society in general or, more narrowly, a professional community. The topic itself was a rejection of the general presumption that animals belong to humans in some vague, perhaps religious, sense. It was a rejection of the notion that as long as they were not being tortured out of some distorted carnal pleasure motive or being innocent recipients of blind violence, animals were ours to eat, entertain ourselves with, wear, and use in science in hopes of improving our lives. These perspectives appeared to ignore the idea that animals had complete and fully meaningful lives of their own independent of the uses made of them by humans, and that depriving them of those lives, or forcibly modifying them significantly, had moral implications.

As we explored the issue of moral status, it became immediately clear that the concept of "personhood" had come to have a profound effect on determining what kinds of beings other than humans "deserved" moral protection and who was left out by virtue of not meeting this conceptual standard. The history of the concept of personhood and the characteristics that came to define a "person" is rooted in a natural focus on human beings, independent of any intended ethical applications. The starting point was that humans are persons by definition and that certain characteristics of that human category are necessary for personhood. The kinds of characteristics that were regularly mentioned in the philosophical literature included rationality and higher-order volition, planning ability, language, autonomy, agency, moral judgment, intelligence, memory, having moral emotions and motivation, awareness of oneself as existing over time, capacity for thoughts and beliefs, and having reasons for one's actions. Consequently, a direct extrapolation of the concept to nonhuman animals implied that the closer an entity appears to be to the normal human being with respect to these characteristics, the greater the ethical protection it deserves. This meant that many minimally cognitive or noncognitive unique characteristics of nonhuman animals were not seen as providing a strong basis for ethical protection. For example, the focus on cognitive criteria left unanalyzed the ethical consequences of housing animals in lab environments that provided few opportunities for the animals to carry out species-typical behavior, behavior described as "instinctive" or "wired in" and not obviously an outgrowth of so-called higher cognitive functions.

In considering this history of the personhood concept, I was beginning to appreciate how limited the concept of harming became when it was dominated by the concept of personhood. I began to wonder whether harming itself needed to be the central focus of ethical consideration, not the personhood or lack thereof of the animal being harmed. While many animals might not share the panoply of higher-order cognitive capabilities characteristic of normal human persons in the metaphysical sense of full personhood, observation of their unrestricted behavior reveals a rich profile of behavioral activities that are carried out with obvious, focused involvement that define in part who they are biologically. These activities, developed over the long course of evolution, include avoiding pain and distress, seeking out grooming partners in social species, exploring the environment, foraging for preferred foods, mating, caring for offspring, defending and expanding territory, building and repairing nests and dens, evading predators, fighting for social status or life preservation, making adjustments to changing climatic conditions, and so on. In other words, animals are exerting control over their lives in myriad ways, and the outcomes clearly matter to them. How could human interference with these behaviors be made ethically irrelevant by the argument that the behaviors don't involve something we recognize as conscious rationality? This made no sense to me.

I already knew that we researchers, at the very least, needed to get beyond the assumption that our own curiosity, well-intentioned intellectual desires, and expressions of sheer ambition, coupled with our ability to easily access and then own animal lives, provided a sufficient ethical warrant to proceed with our experimental manipulations. What intrigued and concerned me was that for some reason the basic economic question of cost, which naturally arises when a human considers ownership of some object or service provided by another, was not reliably asked when it came to animal use in the laboratory. The ethical researcher, I thought, must ask "What does this cost?" and then perform an accurate-as-possible assessment of the costs in harm of using a particular species of animal in a particular experimental context. The extent to which an animal's species-typical behavioral characteristics are blocked or impaired needs to be included as part of the total quantity of estimated harm. Without a sense of the global makeup of the potential harms, making judgments about the balance of costs and benefits can become an empty and biased exercise, the outcome of which is preordained in favor of animal use. I was beginning to appreciate how limiting the evaluation of harm to only those events that constitute the direct manipulation phase of an experiment misses half the point. Assessing the degree of pain and distress at these times is certainly essential, but it is only a part, perhaps even a small part, of a large network of harms that are mostly ignored. That network might include laboratory-induced boredom, disruption of normal circadian rhythms, general anxiety, fear of handling, absence of typical food options, and social deprivation—to mention but a few of the obvious possibilities. How can the costs in harm to experimental animals be appreciated unless we consider these kinds of factors?

In this context, I recalled reading many years earlier about experiments in animal psychology reaching back to the 1920s that used a device called the Columbia Obstruction Box.[2] This device, used to study how strongly rodents were motivated to perform various activities, consisted of three sections: a start chamber where the rat would be initially placed, a goal box that contained some type of stimulus object whose incentive value was being tested (e.g., food, water, a sexually receptive female rat, a newborn rat pup, nesting material, a novel environment), and between them a grid that could be electrified with different levels of shock from mild to highly aversive. The idea was that the strength of the incentive could be evaluated by determining if or how often a rat would cross the electrified grid to gain access to the object visible on the other side. The frequency of crossing and the amount of shock tolerated in order to gain access indicated the motivational importance of the object.

While the researchers at the time would not have put it this way, they had developed, in effect, a procedure whereby they could scale objects and activities that had significant welfare value to the animals. In an ethical context, the method could also have been seen as measuring the degree of harm that was produced by acts like restricting the intake of food and water, depriving animals of social interaction, separating parents from offspring, and so on. In this case, the degree of harm was proportional to the amount of shock tolerated to gain access to the stimulus object. The researchers did not seem to consider this interpretation, only the amount of motivational "drive" that was being created by the various types of deprivation.

The idea of considering a broad array of factors in assessing the harm experienced by an animal also brought to mind the work done by Melinda Novak and Steve Suomi, two of my closest graduate-student friends while I was at the University of Wisconsin. Their oft-referenced paper published in *American Psychologist* in 1988[3] was one of the first attempts to assess the effect of the requirement, codified in the 1985 amendments to the Animal Welfare Act, that researchers make an attempt to minimize harm caused to the "psychological well-being" of animals used in experiments. Their goal was to identify and describe some of the problems the requirement posed for researchers. They expressed concern that compassion-driven anthropomorphism would lead well-meaning researchers astray by uncritically substituting "soft" idealistic notions of well-being for ones that were established through carefully designed controlled experiments. They proposed that well-being is an outgrowth of the interplay between four dimensions: physical health, ability to express species-typical behaviors, absence of physiological indicators of stress, and ability to adequately adjust to environmental challenges. They then evaluated the various welfare positives and negatives of typical lab housing systems used for primates, such as single cages, pair housing in cages, indoor group pens, indoor and outdoor group facilities, and outdoor environments. They pointed out that even if housed in austere conditions like single cages, monkeys could "enjoy" good healthcare monitoring and treatment, would not be subject to direct attack from other monkeys, and could still interact at a safe distance. Furthermore, compared to their wild counterparts, the monkeys would have reduced exposure to infectious disease, live in a more predictable and therefore less stressful environment, and be likely to have fewer problems with degenerative joint disease. The negatives they identified included obesity, lethargy, reduced species-typical behavior, problems coping with change, and displaying stereotyped and other bizarre behaviors.

Then, in a laboratory-research-saving move, Novak and Suomi stated without further explanation that a researcher who was able to ensure acceptable status in two out of the four categories had established a sufficient state of psychological well-being to satisfy the requirements of the regulation. This meant that a rhesus monkey living in a single cage without obvious indicators of disease, exhibiting species-typical self-grooming, showing only occasional stereotyped behaviors, having no evidence of elevated stress hormones, and maintaining appropriate weight and hydration should be considered to be meeting or perhaps even exceeding the definition of psychological well-being, despite receiving daily aversive electric shocks while participating in a learning study. Add to this the regulatory waiver for studies that *intend* to create psychological harm, and in the end you are left with a minimalist standard. It was certainly a sign of improvement that psychological well-being was even being considered, but the AWA requirements were a far cry from taking seriously the idea that holding animals in human-made facilities and interfering with the behaviors they would exhibit in the wild automatically constitutes a level of harm with ethical significance.

Early on, Tom assigned an important article by the University of Michigan philosopher Carl Cohen titled "The Case for the Use of Animals in Biomedical Research." This article, published in 1986 in the prestigious *New England Journal of Medicine*,[4] was at the time (and still to some extent) seen by many biomedical researchers as the definitive ethical position because it wholeheartedly sanctioned the use of animals, and in fact called for increased use. While Cohen agreed that animals deserve decent treatment and shouldn't be exposed to unnecessary pain, it was to him a ludicrous ethical move to entertain any notion of animals' having moral rights that need to be considered in the process of justifying research. I recognized this stance as essentially the same consensus view that had surrounded me since I began my involvement with animal research more than thirty years previously.

I could see as I studied the article that Cohen had a very restrictive position on the definition of a "right" and what kinds of beings could be said to be legitimate holders of rights. His position was strongly related to the traditional personhood positions I had been studying. For Cohen, a right was an absolute trump card that if played by the holder totally blocked any life-altering incursion by another person or group without clear-cut evidence of approval by the rights holder. Further, any previous consent could be terminated at any point by the rights holder without any recourse to some other party. For Cohen, this level of rights power—which allows an individual to overrule arguments for the greater good—correctly belongs only to entities who are members of what he called the "moral community." Members of the moral community are individuals capable of freely making and withdrawing agreements with other members about how mutual treatment would be shaped. This clearly left out animals, who are certainly missing the necessary cognitive capabilities, not to mention the requisite communication skills. Therefore, his position was that when we take over the lives of animals for the purpose of using them in biomedical or behavioral research, we do not violate their rights because they have none.

In my discussions with Tom about the Cohen argument, he pointed out several serious ethical flaws. For example, Cohen ignores an important relationship between having obligations to animals (which Cohen agrees we have) and conferring on them rights (which he argues is not justified). Tom said that what is important here is how a set of obligations come to influence a person's ethical choices. At the individual level, obligations and rights may be disconnected. You can feel obligated to give a homeless person money out of a sense of charity, for example, but that obligation does not then give the recipient of the money the right to demand money again. In contrast, when obligations arise from a societal consensus—such as when a federal law requires researchers to minimize pain and distress in experimental subjects—then it is absolutely coherent to characterize research animals as having a right to be treated in accordance with the requirements of that consensus.

Tom also pointed out the problems with Cohen's assertion that having the cognitive ability to participate in moral communities where members negotiate the terms of mutual treatment and protection and reflect on the coherency of moral beliefs sets humans apart from animals. First, there is an obvious class of exceptions to the rule—many humans lack the requisite cognitive abilities to participate as a morally interested community due to immaturity, brain trauma, catastrophic developmental failure (e.g., anencephaly), or profound neural deterioration (e.g., dementia, permanent vegetative states) and yet are still treated as having rights. In Cohen's system, these individuals maintain the ethical protection afforded to normal rights holders simply by being members of the "kind" (e.g., the human species) that normally is capable of moral agency. Cohen gets these "marginal cases" into the human moral community by arguing that their potential to develop the necessary abilities would have been realized had it not been for an accident of fate. In other words, deserving the ethical protection provided by being a rights holder in the end does not require the individual to be a moral agent, only that she or he be related to the "kind" whose members normally or on average develop into moral agents.

The problem with this reasoning is that once you allow the concept of *potential* into the morally relevant domain, you have great difficulty excluding many animals from having the relevant relationship to normal humans. This is so because primates share with humans a great deal of genetic material, which opens the possibility that technology not yet developed could conceivably alter some aspect of that genetic material to cause the development of the cognitive capacities that Cohen relies on for his human–animal rights distinction. Second, Cohen's declaration that rights are conferred by the presence of moral agency in no way excludes other more basic capabilities, such as consciousness or the ability to feel pain or pleasure, from being considered as additional bases for the conferring of rights. In making moral agency the sole criterion, Cohen is relying solely on tradition.

Since the publication of the Cohen paper, there have been many damaging critiques by many philosophers.[5] Given that situation, it was indeed disheartening to see how many members of the animal research community actively chose to see Cohen's arguments as having settled once and for all the issue of rights for animals.

Another notable experience during the year was taking a course offered by Robert Veatch on the ethical issues in death and dying. Veatch was a wonderful teacher with a strong, authoritative manner and a voluminous mastery of the relevant literature. Likewise, Margaret Little's course on feminist bioethics was an eye-opener. We discussed books like Carol Gilligan's *In a Different Voice*, which emphasized the caring context that surrounds many ethical decisions and what is left out when we seek a singular focus on making ethical decisions from a position of total impartiality. We looked in depth at evidence of discrimination against the study of women in biomedical research. I realized that I must have read about many thousands of animal studies that purposely excluded females because their estrus or menstrual cycles introduced variance that would have interfered with determining the effects of the variables under study. In other words, the tendency to see the male as the true representative of the species distorted basic science and clinical application.

I found that the institute's library on the first floor of the majestic gray and towered Healy Hall, headed then by Doris Goldstein, was truly a place organized for scholarship and discovery. The book, journal, and reprint collection devoted to both philosophical and applied ethics was arranged in such a way that I could hardly walk down an aisle without something worthy of study catching my eye.

But the most valuable parts of the library were the staff librarians, who were scholars in bioethics themselves. They could not only point at but also summarize the important fundamental and new information in virtually any area in bioethics. I hounded people like Martina Darragh and Laura Bishop so frequently that I feared they would start to hide from me. They never did. Instead they showed me how to use the computerized search system and how to subtly adjust my search terms to capture the information I needed. If I preferred, they would do the search for me to ensure that the outcome was the most useful. As a visiting fellow I was allowed to claim one of the comfortable oak carrels in the library as my place of work. The desk that I selected was by a window with a view of the Healy lawn and the campus entrance at Thirty-Seventh and O Streets. I also had a clear view of the statue of the founder of Georgetown University, the Jesuit John Carrol. It was a place where I spent countless hours of discovery and reflection.

*   *   *

During the second semester I took a course in bioethics from Sister Carol Taylor and Edmund Pellegrino. Carol was a nun and practicing nurse with a PhD in philosophy. Ed was an internal medicine doctor, a philosopher, and one of the founders of the discipline of bioethics. He was both a scholar at the Kennedy Institute and the director of his own ethics institute — the Center for Clinical Bioethics. During his long career in academic medicine Ed had been a department chair, a medical school dean, and a president of the Catholic University. (He died on June 13, 2013, the same day he received news that his six-hundredth journal paper had been accepted for publication.)[6] In this class, Ed and Carol emphasized virtue ethics, an approach based on Aristotle's work, and the related idea that applied ethics in the end rests on the good character of the decision maker. They constantly drove home the idea that the clinical provider is a "professional," a practitioner who accepts as lifelong moral obligations the achievement and maintenance of competence, the effacement of self-interest, and caring for the patient as "the one who suffers." The absence of this ethical foundation, they said, makes the practice of providing health care "pure undiluted hypocrisy." They also made it clear that character alone is not adequate for the production of good decisions. Familiarity with ethical theory and an understanding of how to apply ethical principles to particular situations are also required.

Ed held a weekly meeting he called "ethics rounds" where emergent clinical cases in the hospital were presented for discussion and participants worked together to develop an ethical intervention plan. An important aspect of these meetings was that one could not just attend and sit back in a white clinical coat, passively listening to the differing points of view regarding some human medical tragedy. You had to be ready to present your own analysis, position, and justification when Ed's finger pointed at you. There was no comfortable hiding place in the back of the room. This approach was a way to encourage us to embrace ethical analysis as an integral part of clinical and research work, not just an interesting exercise to be indulged in when there was extra time. This was Ed's view of being a professional—someone who strives for a central ethical identity through which research and clinical duties pass on their way to choice and action. His primary teaching strategy was to enact his own personal model of deliberation right in front of you. Applied ethics is composed of three phases, he said: recognizing that an ethical issue is present, analyzing that issue, and creating a plan to intervene with due regard to the individuals involved.

During this semester I also spent two days a week at the NIH Clinical Center in Bethesda, Maryland, serving as an intern on the ethics consult service. Being a patient at the Clinical Center meant participating in the study of experimental treatments that were available only as a last resort to those for whom all established interventions for their disease had failed. This experience, therefore, provided a chance to see the ethical problems that arise when the paths of desperate and dying patients intersect with those of daring clinical researchers wishing to try creative but risky treatments. I saw a number of patients who had originally consented to an experiment but then changed their minds midstream, wanting instead to leave the hospital and die at home surrounded by family and loved ones. Such decisions often conflicted with the goals of the researchers, who urgently sought to learn whether the experimental treatment in question was of any value. Sometimes the researcher was unable to understand or accept the patient's decision to let life go.

In one case that stands out in my mind, I saw a researcher actually call security in hopes of stopping a patient who had changed her mind from leaving the hospital. After the shocked officer pointed out that he could intervene only if the patient was stealing government property, the researcher called in the ethics committee, hoping we would find the person incompetent. She was clearly not incompetent. The patient understood well the consequences of leaving the hospital and withdrawing from the protocol. She told me that her desperate desire to live had finally calmed and that the brutal side effects of the continued experimental treatments were now destroying what little was left of her life. She wanted only palliative care and the chance to share her last moments with her friends, family, and loyal dog Irene. It was this last wish that left one of the researchers still insisting that she was incompetent. Give up a 1-in-100,000 chance to receive a clinical benefit from an experimental treatment to be with her dog? The wish did not seem at all alien to me. I tried to articulate the sanity and meaning of this relationship, to explain that Irene was an essential filament of the web of her family.

The next step taken by the researcher was to call the woman's husband. He arrived, and the two went back to her room to talk. Standing by the nurses' station, I could hear the "talk." Anyone who wasn't stone deaf could hear the husband's voice: "You owe this try to the family; you are going to stay" was the gist. He came out of the room and delivered the "good news" to the treatment team: "She is staying." The lead ethics consultant asked for a chance to talk with her. The husband stood in the threshold, looking like a bouncer at a nightclub. "No one is talking to her." The ethicist backed off. To push the issue at that moment would just inflame the situation.

The woman died about a week later, in her room at the center, surrounded by her caring nurses. Her family members, including Irene, were back in their home in Virginia.

Thus at times a Clinical Center researcher's passionate desire to discover knowledge and improve the lot of future patients obscured recognition of a current patient's pain, distress, and mortal exhaustion. I saw support for Henry Beecher's statement in his classic 1966 paper on the ethical problems in clinical research: "Doctors are dedicated to their patients. Researchers are dedicated to their protocols."[7]

I also saw many patients who, while recognizing that their participation in some of the experiments offered little or no hope of clinical benefit for them, still courageously offered up their lives to advance the possibility of cures for others. I saw how entering into those studies could add tremendously to a person's sense of value and pride, joining his or her life to the lives of others who would face the same medical predicament. I was reminded of a lecture I had attended years earlier in which the psychologist Allan Wheelis discussed the impact of Eastern religions on modern psychological concepts. The Buddhist concept he found most enlightening was the idea that our external skin is an "arbitrary boundary." Watching the transformations that some of the research volunteers experienced, I could see now what he meant.

Nonetheless, the center was also a place where fear of death, a wish to discover and help, ambition, and misinformed consent could impel some ethically insensitive researchers to exploit patients. The worst offenders in this small group earned the designation "cowboy" from the ethics team. I could clearly see how a similar dynamic had worked in my own research life. The desire to know had eclipsed the appreciation of what was being asked of the animals.

Barbara and I continued to meet independently of the tutorial with Tom. She had recently published a book on animal research titled *In the Name of Science* that became a focus of our discussions.[8] It was a well-reviewed book covering the history of animal research from its roots in the animal dissections of Galen in the second century to current controversies and ethical problems. In spite of the acclaim, she lamented that the book lacked some of the passionate skepticism she had developed over the years about the level of care routinely available to laboratory animals and the usefulness of a good deal of animal research. She felt that this material had been chipped out during the editorial process. She said that her publisher, Oxford University Press, surprisingly had selected an official of a one-dimensional pro-animal-research group as one of the initial reviewers of the unpublished manuscript. As could be predicted, that review of the original manuscript was harshly negative. In order to stay with Oxford, she removed much of the material that had been deemed too radical. She deeply regretted her decision to "cut the ethical heart out" of the book.

Barbara began to enlighten me about the importance of developing working relationships with some animal protection groups as well as with scientists. She then facilitated the beginning of one such relationship when we went together to discuss animal welfare issues with her own primary professional model, Christine Stevens. Christine had founded the Animal Welfare Institute (AWI) in 1951 to promote, among many broad animal-welfare issues, improved treatment of laboratory animals, not the abolition of animal research. Christine had been influenced by her father, Dr. Robert Gesell, who was chair of the Department of Physiology at the University of Michigan and had long spoken out about the need for more humane treatment of the animals used in physiological research. In 1952, at the business meeting of the American Physiological Society, he openly lamented that some of the research being carried out by researchers was a "travesty of humanity." This statement, and others based on his own experience as a researcher, led to attacks from many colleagues and claims that his concern was symptomatic of a neurological impairment.[9] Despite consistent resistance from portions of the animal research establishment, the AWI survived and prospered.

In 1994 the AWI resided in a large and slightly worn Georgetown mansion not far from the Reservoir Road entrance to Georgetown University. The first time Barbara and I visited we were greeted at the fence by a number of friendly, ragged mixed-breed dogs, one with three legs and all with wildly wagging tails. As we entered the house, the staff was preparing for the daily group lunch around a large dining-room table beneath an ornate chandelier. We were asked to join them. Once seated, Christine introduced me to the group; Barbara was already well known to all. A simple vegetable soup was served, and plates of steamed vegetables and bread were passed around as the conversation began. I recall that the discussion toured the topics of state laws regulating the use of leg-hold traps, stressful housing conditions for Canadian horses used in production and collection of pregnant mare urine, which yields estrogen, the basic ingredient in medications for postmenopausal symptoms, and the development of the next edition of the institute's book, *Comfortable Quarters for Research Animals*. This publication encapsulated the institute's goal for animal research: to provide enriched and species-tuned living environments for lab animals for the sake of the animals and the benefit of useful research outcomes. For the most part Cathy Liss, a young woman with a strong presence, led the discussions. Christine chimed in from time to time, often providing historical context to the issues.

After lunch, as the others returned to their work, Barbara, Christine, and I sat together. I found Christine to be quite mesmerizing. Although late in middle age and dressed very casually, she was an imposing presence. She spoke with clarity and authority. She asked what had motivated me to come to Washington and how I was going to use my limited time for the cause of animal welfare. She reminded me of Ed Pellegrino when he pointed his finger at someone during ethics rounds. She seemed authentically interested in hearing my story, but not just out of curiosity. The question "what are you going to do for animal welfare?" hung in the air as encouragement but also as a challenge. She was clearly asking me not to limit my study to the philosophical theories of animal moral standing without formulating an action plan to improve laboratory conditions for animals. She emphasized that my animal research background provided an important platform from which to work.

This notion was made clearer to me a few weeks later when we both attended a conference at the Institute of Medicine on lab animal welfare. During one talk, a presenter was pressing what he saw as the absurdity of regulatory requirements for the treatment of laboratory rodents. While we poison and trap the wild rodents that enter our homes, he noted, federal regulations require researchers to provide very similar rodents with clean and roomy cages, nutritious food, expert veterinary care, and pain-free deaths.

During the Q&A period I asked the speaker if he saw a similar absurdity in the contrast between the military bombing strategies that targeted civilian populations during World War II and Vietnam and the requirements for protecting humans participating in biomedical research. He was unprepared for the question and went mute. I insisted that if more people were aware of the extent of pain perception in rodents, people might call for different methods of dispatch.

At the end of the conference I went to greet Christine and Cathy. As we parted, Christine turned to me and said, "That was a nice speech, John," and walked on. What I heard was not an affirmative judgment but an implicit question: "Are you going to do anything more than give an anonymous little three-minute speech to a tired and inattentive audience?" It was a very good question.

After the conference I met with the leaders of other animal welfare organizations, including Andrew Rowan and Jonathan Balcombe of the Humane Society of the United States, Tina Nelson of the American Anti-Vivisection Society, and Ken Shapiro of Psychologists for the Ethical Treatment of Animals (PsyEta). I came away from these meetings with a different view of the animal welfare movement from the one I had previously, which was based mostly on the views of my research colleagues. That picture was of groups of raging fanatics with distorted emotional capacities, ignorant of science and medical history, who saw all researchers as sadistic boobs. Instead what I experienced were people morally troubled by the suffering experienced by animals in their interactions with humans, knowledgeable about science, and devoted to effecting change. For example, Ken Shapiro wanted deeply to engage researchers in long conversations about research, values, and alternatives to the use of animals — not to condemn them from a distance. One other characteristic was clear: a willingness to adjust their lives to accommodate their concerns and ethical beliefs. This included the way they planned their diets, what they wore, and even their expectations for health care.

I came away wanting to work with these groups in the future. I thought that their approach had a real chance of improving the conditions of lab animals. Unfortunately for me, the collaboration would not materialize to the extent that I desired. Perhaps I was still too close to the animal research community to be completely trusted. However, many years thereafter, when my commitment to the cause had matured, Ken Shapiro would ask me to participate in his organization, serving as reviewer for his journals and later as an advisor to the Animals and Society Institute, an organization that he founded.

Besides engaging us in the study of ethical theory and bioethics and exposing us to dilemmas of treatment and research that occurred at the bedside and at the end of life, our mentors at the institute encouraged us to discuss the moral standing of animals in open debate. They judged this activity to be a necessity of a serious moral life. My experience up to this point had convinced me that such discussions had to be undertaken with great care and only in a few select academic settings. If you claimed that animals possessed rights, your view was likely to be held against you by some members of the research establishment. For many, it was simply heresy.

At Georgetown, the debate about animal rights was welcomed because it exposed the incoherence of our values regarding inflicting pain for our primary benefit on beings who had the capacity to experience pain but lacked the ability to resist. It was also a feared discussion, because the conversation topics often went beyond research uses of animals to the consumption of animals as food and their role as entertainers and personal companions. I often observed that if a lunch break followed one of these debates, the meal choices of the fellows tended to move to salads and tofu in lieu of chicken and roast beef.

*　　*　　*

As the transformative year drew to a close, my imminent return to New Mexico brought with it a central decision: would I be willing to take the job of department chair as I had discussed with the dean before I left for Georgetown? While I had promised only to consider the position if it were available when I returned, there had been a vague but real sense in our discussion that I would more than likely take the job.

I dreaded having to make the decision. My desire to promote ethical advocacy in animal research and research in general had grown in intensity. This was what I really wanted to do with what was left of my professional life. While being chair might provide a platform for that advocacy, it also involved class schedules, teaching assignments, dealing with interpersonal squabbles, and endless committee work at all levels of the university and community. There was no way that the important work of a department chair would not interfere with my plans. Not being chair would give me the flexibility to start a research ethics consulting service, an idea that had begun to develop. I could also expand my teaching to more research ethics courses, and do so more competently than I had in the past. So this was the choice: be department chair and use the opportunity to develop my skills as an administrator and move through the ranks, perhaps to dean, or focus on ethical advocacy through teaching, research, and direct service.

The call from the dean finally came in late October, a little less than two months from my planned return to New Mexico. When I picked up the two-day-old message slip at the institute's office and read the caller's name, there was no doubt about its purpose: decision time had arrived. I proceeded to a meeting that I had scheduled with Barbara. I thought that as the day passed I would call Charlene and talk with my closest friends among the fellows, Henrik Jørgenson, a Danish philosopher, and Diann Uustall, a doctoral-degree nurse, and from those conversations my final response would develop.

At our meeting Barbara had two issues to discuss. The first was that she planned to hold an animal ethics conference at Georgetown at the end of August and wanted me to give a presentation on the transformation in which I was involved. Second, she said that she and Tom wanted me to join a group of authors who were developing a volume for Oxford University Press with the working title *The Human Use of Animals*. Both invitations were unexpected, and I saw them as expressions of my development during the fellowship. Participation in these projects would ensure some continuity with the people and the work I had been doing at Georgetown. This was important, because returning from sabbatical requires some active resistance to familiar expectations and needs that draw one back onto the well-worn path trod previously. Perhaps these plans would contribute to my ethics focus even if I took the job as chair. I immediately accepted both invitations.

As the meeting was coming to a close and I was packing my papers, the institute's secretary came over to deliver another message from the dean to call; this time it was marked urgent. "He sounded annoyed," she added.

Instead of going to the office to make the call, I decided to go to my sacred desk in the bioethics library in Healy Hall. As I walked, I found that my step had a bit of swing to it, put there by the invitations from Barbara and Tom. I rehearsed again the implications of being chair and what I might be able to accomplish. The year that I had spent as interim chair had been very hard and conflicted. On the other hand, I thought I was effective in working with the problems of undergraduate and graduate students. I also like helping new faculty get started, as Frank Logan had once done for me. Maybe it was time for me to prove to myself that underneath my frequent complaints about administrative decisions was an ability to improve on what I had seen. Then there were the substantial financial benefits that went with the post; they would change the curve of my salary for as long as I stayed at UNM and would boost my retirement income. The dean was counting on my taking the job and had financially supported my time at Georgetown. This last issue carried weight. I had accepted the money knowing what his expectations were. When I got up from my desk to return the phone call, I had mostly decided to accept the offer if he made it.

As I started to walk through Healy Hall toward the pay phone near the spiral steps, I took a detour. I stepped out onto the patio in front of Dahlgren Chapel. Off to the right was Old North, the central building of the university until 1859. From its porch twelve presidents of the United States had addressed the students, starting with George Washington in 1797. I sat down on a bench facing the fountain in the foreground of the courtyard leading to the always-open-for-meditation chapel. Signs of a brutally cold winter to come were everywhere. The trees were dropping their leaves as if to pull their energy in closer to their cores. There was a bite in the breeze, and clouds gathered dark and thick. Birdsong was gone from this tree-filled place, and the squirrels moved with new haste. Nature was preparing for change. What kind of change was I preparing for?

This moment felt very different from the send-off party at Elaine Moran's house in August 1971, before I left for the job in New Mexico following graduation from the University of Wisconsin. It also felt very different from the time when I completed my clinical fellowship at the University of Washington. In both those cases I was bursting with enthusiastic certainty. I remembered telling Dr. Arnie Katz, who provided my required training therapy during my clinical fellowship at the University of Washington, that as I prepared to return to New Mexico I had an image of myself as full of warm light that radiated outside my bodily boundaries. I had been ready and excited to teach about the new perspectives and skills that I had acquired. Now, as I sat in the chill between Healy Hall and the chapel, I was feeling a lot less comfortable. The year of classes, discussions, reading, and direct experiences in labs and hospitals had left me excited but also feeling something heavier, as if I was shouldering a new obligation, a mission, a trust. Perhaps what I was experiencing this time was the "calling" that Ed Pellegrino always talked about when he lectured about what it means to be a true professional.

Why then was I interested in being chair? How did this motivation compare with my other pull—to expand ethical mindfulness among developing researchers so as to counteract their exposure to many mature researchers' indifference to the ethical costs of animal and human use? Ethics and science had been traditionally separated. As a result, the values of discovery, productivity, and achievement were permitted to dominate research motivation, leaving it unchallenged by awareness of the human and animal costs. I had come to undersatnd that no activity that affects the welfare of humans or animals should be free of ethical reflection. Ethics is above all a system whose purpose is to remind us to consider the impact of our choices in a larger context than just ourselves and our presumed lofty goals.

At similar decision points in my professional life I had consulted the people responsible for my graduate school education—people like Harry Harlow and Jim Sackett—for advice. When I was considering pursuing a fellowship in clinical psychology in 1976, both offered their encouragement and help. Now I wondered what they would tell me. I was pretty sure from watching Jim's own career what he would say: "Stay away from administration."

Imagining Harlow's response was more complicated. Harry had become known as a prime example of the cruel and thoughtless animal researcher. Nonetheless, I was not so sure that he would have thought bioethics a foolish career direction for me. He was one of the first American psychologists to emphasize the existence of a broad and powerful emotional and cognitive life in rhesus monkeys. He was the one who promoted the idea that monkeys made decisions, tested hypotheses to solve learning problems, and had social relationships grounded in affection and love. It was my recognition of these very characteristics of my monkeys that had raised the question whether my research could be morally justified. The fact that Harry hadn't made the ethical connections didn't mean he would have disrespected someone else who did make them. If Harlow modeled anything consistently to his students, it was intellectual independence. He encouraged us—no, dared us—to think freely. As students we were not burdened by duties to do "his" research projects. There was nothing to stop any of us from choosing another ethical path, one that diverged from the devastatingly harsh methods he had adopted. There is no doubt in my mind that if I had still been a student in 1975 and had asked if I could use his grant money to buy a copy of Peter Singer's newly published book *Animal Liberation*, in which some of his work was criticized as cruel and useless, so that I could study the relevance of Singer's ideas, he would have immediately approved the purchase. I also believe that he would have said, "Get two additional copies; one for me and one for the library."

The fact that such a conversation had not taken place with me or with anyone else, as far as I know, was why I was having the dilemma that I was having. Conversations like these had been absent during my own education as a scientist, and they remained so for most young scientists of the day. I was now in a position to help improve that situation.

I walked back into Healy, went directly to the public phone, and placed the call. Luckily, the dean was available. He immediately offered me the position of chair of the Department of Psychology. He added that the majority of the faculty supported my appointment.

I waited for the first open space in the conversation and began to explain why I was not going to accept his offer. I described my idea to develop an academic unit, the Research Ethics Service Project, that could promote research ethics education and serve as a consultation service to resolve ethical conflicts.

He expressed disappointment in my decision but did not try to talk me out of it. I think he could sense the earnestness that underlay my decision. At the end of the conversation, he offered that he thought the project I was describing would work best if it were situated in the office of the vice president for research. I thought this was very impressive. And it indicated that his frustration with me did not cloud his judgment about the potential importance of such an endeavor. I hadn't expected to be buoyed by this conversation, but I was.

The rest of my time at Georgetown sped by. While I began to feel an urgency to begin my new career, I also began to anticipate the loss of the daily encounters with my friends. Henrik and his wonderful wife Marianne would go back to Denmark; Diann would return to her practice in Rhode Island; Jochim Vollman would soon leave for Germany. The rich interactions we'd had at lunches and dinners, so full of reflection and laughter, would become relegated to what could be squeezed into occasional phone calls and e-mails. But I would be home with Charlene, Jay, and Katie. Charlene would have to become my primary local ethics discussant. Luckily, she seemed to have quite an aptitude for the topic. On many occasions during our evening phone conversations I would describe to her some concept I had been struggling with. Almost without exception she would grasp the subtleties and end up explaining it to me. I was, however, confident that Tom Beauchamp, Barbara Orlans, and I would remain in close contact. We understood one another's experience and motivations, and we had just agreed to work together on two projects.

There was one other friend I needed to say goodbye to, and I kept my eye out for him as I walked the streets around the Georgetown campus. He was a large black and gray cat that I'd first met eight or nine months earlier on what are referred to as the Exorcist Steps—a set of steep stairs that had appeared in a famous death scene in the 1973 film *The Exorcist*. On most evenings as I descended these stairs on my way back to my apartment across the Key Bridge, I would encounter the cat sitting amid the twisted ivy at the top of the stairs. Most times he accepted a scratch, rubbing his large head firmly into my hand and the inside of my forearm. Other times he seemed just casually interested, permitting only fleeting physical contact. I hoped that he lived in one of the upscale Georgetown townhouses in the area or perhaps cohabitated with a student in more modest accommodations. In either case, his coat always appeared combed and his ribs were padded. Our encounters were sufficiently frequent that I looked forward to the possibility of seeing him each day as I headed home.

Just east of the Key Bridge, I entered the small community park at the north end of the bridge. It was dark, and the surrounding street-lights shimmered in the cold, clear night. Hearing a meow, I looked down; at my feet my friend was emerging from the underbrush by the sidewalk. "How did you get down here?" I asked with initial delight and then foreboding as I watched the car traffic rush east and west on M Street and south and north on the bridge. I stroked him several times as he curled his body around my leg, rubbing back and forth.

Then I followed him as he marched to the west-facing curb. He stared intently at the pattern of Key Bridge traffic with its glare and noise. I found myself imploring him, "Don't do it, don't do it," as he clearly appeared to be preparing himself to dash across the six lanes of rapidly moving bidirectional traffic. I wanted to grab him and protect him from the choice he seemed ready to make. I hesitated, though, because I was unsure how he would react to being picked up, and he frankly looked like he knew what he was doing. With the traffic flowing, he dashed out, his tail held high, with the speed of a sprinter hearing the starting gun. He made a couple of lightning adjustments to a slightly semicircular course and was then gone from sight. In fear I looked for evidence of his mangled body on the road-way. There was no way he could have made the crossing unscathed. Then I saw him sitting on the sidewalk directly across from me. He turned around and headed in the direction of the Exorcist Stairs. Getting to them would require yet another street crossing, but this time with only three lanes of traffic. After what I had just seen, I was confident he would make it.

As I continued my walk home I barely noticed the brutally cold wind that blew from east to west. I was engrossed in wondering how the cat was able to accomplish the maneuver I had just seen. There was so much that he had to deal with simultaneously: the control of fear; the speeds of the cars coming from two directions and multiple lanes at the time of takeoff; the changes in automobiles' speeds as he ran. It was incredible. Nothing that I knew about cats provided me any base from which to understand the lightning-fast decision-making skills that I had just seen.

It would soon be my turn, I thought, to negotiate a similarly dangerous thoroughfare—the one trafficked by animal researchers completely ignorant of, and unwilling to see, the kind of cognitive magnificence my feline friend had just demonstrated. I hoped that I shared enough of his determination and mental agility to get to the other side.

Leaving Georgetown to return to New Mexico was bittersweet. I had an early lunch with Barbara Orlans, her husband Herb Morton, and Diann Uustall at Diann's apartment, which was just a bit south of the National Cathedral. Throughout the meal the chimes from the cathedral played fitting themes that at times seemed appropriate for a wedding and at other times a funeral. At the airport it seemed that the cosmos was also ambivalent, as my flight kept getting delayed, first because of equipment failure, then emergency help for an ill passenger, and finally thunderstorms. As we sat on the tarmac wait-ing for the last storm to clear the vicinity of Washington National Airport, I did not feel annoyed about the holdup. It was as though a little extra time in Washington might allow me to come up with a way to stay, keep studying, and avoid the responsibilities waiting for me in Albuquerque.

7

Protection

*How easy it is, Doctor, to be a philosopher on
paper, and how difficult in real life!*

ANTON CHEKHOV

It is better to be in harmony with oneself and at odds with others.

SOCRATES

On the flight home from Washington, DC, I started to sketch out the
form of the Research Ethics Service Project (RESP) that I wanted
to create at the University of New Mexico. By the time the plane got
to the Dallas–Fort Worth Airport to make the Albuquerque connection, I had outlined an ambitious proposal that was composed
of four elements. To fulfill an educational function, the unit would
offer both stand-alone presentations on human and animal research
ethics and a full semester course with both content areas covered.
A speakers' bureau would be available for presentations at public
venues. To help the IRB and IACUC committees resolve difficulties, the unit would also have a consulting function, and this would
include a mechanism for professors, students, and staff members
to submit ethical questions (anonymously, if desired) and receive
detailed answers. Finally, the unit would organize a university-wide
ethics lecture series hosting speakers from outside the university to
talk about topics of general interest. Although the overall goal of the
RESP would be to increase ethical literacy, I also had in mind the

specific goal of working systematically to expand awareness of the ethical issues arising from using animals in research and teaching.

I had already decided that if I could get the project going I would invite Dr. Patricia Heberer of the United States Holocaust Memorial Museum in Washington, DC, as the first speaker in the lecture series. Barbara Orlans and I had recently spent a day with Dr. Heberer discussing Nazi science, and we were both fascinated by the issues and her brilliant scholarship. The process whereby a country like Germany, with the most progressive medical system in the world, could be so horribly transformed had far more than just historical relevance. I thought that general interest in Nazi science and the gargantuan moral failures that it exemplified would serve as a powerful draw to faculty, students, and the general public. The next speakers would come from among the group of scholars that had just provided me with such transformative experiences, such as my mentor Tom Beauchamp. Just thinking about these possibilities was exhilarating.

By the time I got to Albuquerque, I had decided to stay with the name Research Ethics Service Project. This wording was intended to communicate that the program was neither the "ethics police" nor a formalized system of required oversight or compliance, but rather an entity designed to provide assistance when ethical issues emerged.

I was confident that there would be interest in the RESP, in part because a number of high-profile scientific misconduct cases had been prominently featured in the news over the previous couple of years. The Sloan Kettering cancer researcher William Summerlin, the Nobel laureate and noted HIV researcher David Baltimore, the rising-star cardiology researcher John Darsee at Harvard University, and Stephen Breuning, a national authority on the effects of tranquilizers on the developmentally disabled, had all been either charged with fabricating data or found to have fabricated a substantial amount of data in their published research. Most striking was the fact that all of these men worked on highly visible and important clinical topics, not arcane niches. In addition, these quite visible cases had caught the attention of federal oversight committees, and the resulting public hearings had stimulated the production of a program titled *Do Scientists Cheat?* that appeared on public television across the country in prime time. The disturbing answer to the question in the title was yes, and it was not a modern development.

Once back in my university office, I immediately began to expand the outline into a full packet of information replete with course and presentation outlines, a list of potential invited outside speakers, a survey of courses at UNM that taught research methodology involving humans and animals and would benefit from an ethics component, and a limited budget that would primarily cover half my salary. I quickly made an appointment with Dr. Ellen H. Goldberg, the vice president of research and an outstanding researcher whose work on cell-mediated immunity was highly respected. I asked for a half-hour appointment and practiced my presentation so that it would fit into that amount of time. It would be my first chance to find out how the activity I now saw as my calling would fit into my home academic environment.

My plan was to start the presentation with a review of the stunning article published in 1993 by the highly regarded sociologist of science Judith Swazey[1] in the *American Scientist*. Her study surveyed the rate of exposure by graduate students and faculty from departments of sociology, civil engineering, chemistry, and microbiology to acts of inappropriate data manipulation, sloppy lab practices, violations of research regulations concerning the treatment of human and animal subjects, data fabrication and falsification, and plagiarism. The results showed that none of these acts was rare and some, such as plagiarism, were fairly common. Perhaps most disturbing was the finding that nearly 50 percent of the students and 26 percent of the faculty members surveyed feared they would suffer retaliation from supervisors or colleagues if they dared to exercise their ethical responsibility by becoming whistle-blowers. I planned to present to Dr. Goldberg two other cases as well. The first was an investigative report by Eileen Welsome in the *Albuquerque Tribune* (November 1993) that revealed for the first time the existence of experiments where hospitalized patients were injected with plutonium so researchers could study the way the substance gets distributed throughout the human body. The second was what had become known as the "Silver

Spring monkey case." A leading neuroscientist, Edward Taub, was studying whether the movement disabilities produced by cutting the sensory nerves in the arms of macaque monkeys could be rehabilitated by requiring them to use the nonfeeling arm. An undercover operation had made public photographic evidence that was readily judged to depict grotesque and inhumane treatment of the monkey subjects.[2] Prior to this revelation, Taub had received nine consecutive years of financial support for this research from the NIH.[3] The matter of the ethical treatment of the research monkeys had been in the courts since 1981 and still raged as a point of contention between many neuroscientists and the animal protection movement.

In my mind, these three cases provided prima facie evidence that there was significant ethical disarray in many research laboratories across many academic disciplines—basic science, social science, and biomedical research—and both human and animal subjects were affected. These findings challenged the traditional idea that professional ethics was best taught "at the bench" through a form of intellectual osmosis. Rather, it appeared that this community-responsibility approach left no one accountable for determining the ethical knowledge and attentiveness of student scientists. Many teachers felt that they were just too busy or unable to deal with ethical issues, and so they were treated, if at all, as afterthoughts. The approach needed to be at least supplemented by a real curriculum that included exposure to ethical theory, relevant codes of conduct, and discipline-related case studies. Ethics needed to be presented as a real and substantial discipline, not a collection of mere opinions where each one was deemed as good as the next. In my meeting with Dr. Goldberg, after making this basic point about the need for RESP, I would transition into its proposed structure and the budgetary requirements for initiating the program.

When I met with Vice President Goldberg, I began by handing her an outline of my discussion plan. As soon as she finished scanning the document and looked up, I started to speak. I felt an incredible urgency but wanted to appear relaxed, so I talked very deliberately. I could see that the Judith Swazey data caught her attention right away, and she was obviously familiar with the radiation study controversy.

However, once I commenced presenting the introductory details of the Silver Spring monkey case, she stopped me in midsentence. Evidently Dr Goldberg did not wish to hear any more details about monkeys with open sores on their arms, missing fingers, and the ethical failures of respected neuroscientists in conflict with the animal rights movement. She had two questions: Did any of our "peer institutions" (e.g., public universities in Arizona, Colorado, Utah) have formal research ethics and integrity programs? What was the evidence that ethical behavior can be taught?

I responded that I didn't know if our peer institutions had programs, but Harvard had created the University Center for Ethics and the Professions with political philosopher Dennis Thompson as the founding director eight years earlier, in 1987. I told her that I had just returned from a fellowship in bioethics at the Kennedy Institute of Ethics, which had been established at Georgetown University in 1971. As for the profound question of whether ethics could be taught, I conceded that "formal education will rarely improve the character of a scoundrel," loosely quoting Derek Bok's classic paper on the subject,[4] but said that I was certain that most students could be taught to sharpen their ability to detect when an ethics conflict was present and also could learn how to collect information relevant to reasoning through to a set of ethically acceptable choices. I explained how I thought practicing ethical analysis on cases relevant to a student's research domain could lead them to have some confidence in their ability to effect change during a "real" conflict. Besides, the work would help to identify those students who showed interest in gaining some facility in working through ethical conflicts and those who were deficient or at worst indifferent. Armed with this information, mentors could decide whether closer supervision was required or increased trust should be granted. In any event, the focus on ethical issues would communicate that having ethical knowledge and facility was part of being a professional. I closed by saying that since I was an active researcher myself, this might lend my program a degree of credibility not immediately available to nonscientist "outsiders."

Dr. Goldberg seemed only mildly interested in my proposal and said that I would need to collect more information before she could make a financial support decision. This would turn out to be a more positive attitude than what I would experience from the upper administration in the coming years. Of the five vice presidents for research with whom I would work directly on this issue, only one, an engineer named Dr. Nasir Ahmed, who held the position briefly as an interim, would show significant principle-based interest. Others appeared openly resentful about the resources the RESP program required. One vice president, asked to give the opening welcome to a university-wide conference on research ethics, started his introduction by saying, "Let's face it, research-related ethics and compliance matters are a pain in the butt."

Luckily for the RESP, Dr. Ahmed was placed in charge of the research office after Dr. Goldberg left to take the prestigious directorship of the Santa Fe Institute. During one of our early meetings, I was surprised when he readily authorized the money and gave the go-ahead to begin putting in place a significant part of my comprehensive proposal. He also asked that we formally survey the faculty and graduate students about whether they saw a need for improved ethics education and what form it should take. Importantly, his purpose in this was not to determine whether we should go forward with RESP but rather to inform us about the necessity to stimulate interest if it turned out to be lacking and about what content material we needed to present. Vice President Ahmed also reappointed me chair of the Institutional Animal Care and Use Committee (IACUC) and a member of the human research Institutional Review Board (IRB). He said that my familiarity with ethics issues justified the appointments. He clearly believed that ethics is as important as creativity and scientific competence to the research process.

* * *

Over the years of RESP's existence, I regularly received requests to discuss the ethical dimensions of a number of biomedical and healthcare issues, including keeping patients in vegetative states alive, refusal of treatment, physician-assisted dying, and abortion. In terms of human research, there was considerable interest in lectures on the use of deception and placebos, what makes informed consent valid, the ethical justification of research regulation, and scientific misconduct issues like data fabrication and plagiarism. In contrast, faculty of many departments in arts and sciences, engineering, and health sciences, while very interested in the broad topic of bioethics, expressed little or no interest in having a talk strictly about animal research ethics.

There were occasional requests for on-campus debates on animal-use ethics, often at informal lunchtime meetings of ethics interest groups, where relatively few students would be present. Usually I would be paired against one of the university's noted animal researchers. On one of those rare occasions, I was asked to debate a physiologist who believed that random-source dogs (i.e., unclaimed dogs retrieved from city shelters) were absolutely necessary for use in cardiac physiology labs for medical students. I was expected to argue against their use. A small but strident group of medical students had stimulated interest in the issue by raising questions about the necessity of the animal labs. I knew of the physiologist's work and had been involved in deliberations about these same labs when I was on the medical campus IACUC several years earlier (see chapter 5). Frankly, I was surprised and disappointed to hear that the controversy had not been resolved years earlier.

When the debate was supposed to begin, I did not see my opposition in the room and thought that perhaps he was late or that another colleague would be speaking instead. However, after his introduction by the coordinator of the lunchtime lecture series, he jumped out of the audience like a clown out of a birthday cake and bounced energetically up to the lectern, alternately waving and pointing to a bright white T-shirt with large black letters and a red heart that read "I love my Labrador." As he had no doubt hoped, the group laughed heartily. He then pulled the T-shirt over his head and wore it for the remainder of the discussion. It became clear to me, once he began to speak more formally, why he was a favorite among many medical students. He was animated, articulate, and oozed absolute positional certitude. His opening remarks were blunt in their unnuanced rejection of "animal rights" as an absurd notion. He referenced Carl

Cohen's argument that moral standing at the level of having rights requires membership in a community of morally capable agents, which leaves animals permanently out. He did not waste any additional time defining his terms or elaborating on the logical rationale for his and Cohen's dismissal of animals' moral standing. This apparently wasn't necessary; his authority as a full professor with many peer-reviewed publications was sufficient. While his T-shirt was intended to make it clear to the audience that he was not an uncaring person when it came to animals, he forcefully defended the use of dogs—other than his own—in teaching cardiac physiology, asserting that there is no other way to illustrate the dynamic changes and interactions that are intrinsic to that physiological system. Besides, the animals used were about to be euthanized anyway, and using them for this purpose made their lives more significant. That is, significant to humans.

He rejected completely the value of computer-generated models because they could in no way capture the spontaneity and intrinsic variability of real animal-generated data, and besides, they did not provide the student with the "feel of flesh" that was also a necessary part of the experience. He pointed out that after surgically cracking open the dog's chest, the students are able to touch the beating heart and feel the lungs, not just stare at a computer screen. I recalled the feel-of-flesh argument from the IACUC deliberations of the past, as well as my observation that students in the lab had had to hack and burn through the chests of their underanesthetized dogs with soldering irons instead of cauterizing surgical units.

The professor finished and began to walk away from the podium but then hesitated and returned to add one more comment. He said that we should stand up for the sacred principle of academic freedom, that an instructor "had a right" to teach the way he or she knew or at least believed to be effective—end of discussion. The position was clear: the decision to use animal lives in professional education was at our total discretion.

When it was my turn, time was already short. I tried to establish a new atmosphere that was not permeated by glibness. I spoke of ethical complexity, uncertainty, and ambivalence. I contended that the moral status of animals had been debated for many centuries and that our discussion today was an extension of a persistent vexing question. I showed slides of the representations of animals found in ice age caves of northern Spain and southern France and talked about how aboriginal hunters apologized to the animals they hunted and killed. I mentioned the edicts of Emperor Asoka of India, who, upon his conversion to Buddhism in the third century BC, attempted to reduce the consumption of animals as food and to eliminate their use in sport fighting. I quickly highlighted a few of the incredibly inconsistent ways that animals were treated in modern Western culture—billions of dollars spent on domestic animal care and tens of thousands of animals tossed away like just so much Styrofoam trash. I wanted to point out that these paradoxes were evidence of a powerful ambivalence about what animals deserve in terms of ethical protection, that this ambivalence and treatment incoherence indicated a powerful ethical conflict that ought not be crushed or dismissed but should motivate a deeper understanding of the ethical issues involved.

I then moved to the specific focus of the discussion and summarized the data on whether animal alternatives are capable of teaching important concepts like those found in cardiac physiology labs. Indeed the data available supported the conclusion that computer simulations were capable of facilitating deep learning in students, and that students so instructed often performed better than those more traditionally trained with dog models. I also laid out conclusive evidence that using animals was a serious distraction to a sizable minority of the students, who objected to their use on moral and ethical grounds.[5] This moral objection required respect and should not just be bulldozed like some insignificant personal preference. In fact, I said, these students were trying to make a connection between their moral intuitions and their professional conduct. I concluded by saying that we have a duty to reduce the harming of animals that are clearly capable of pain and suffering, especially when successful alternatives are available.

During the brief Q&A session that followed, the professor was asked whether he permitted students who conscientiously objected to harming animals as part of their education to learn the information differently. His face contorted into a stage grimace, and he dramatically growled like a pirate captain while saying that he "tried" to work with such students. He must not have tried too hard, as this debate had been organized in part due to students' complaints that there no alternatives were available to them. After observing his demeanor, I thought that a student looking for alternatives to live dogs would have to be strong indeed to defy this man's confidence and forceful temperament.

I interjected a few comments about the case of Safia Rubaii, a medical student at the University of Colorado School of Medicine who in 1992 refused to participate in a required cardiac dog lab but passed all the exams and was still failed and denied promotion to the next year. This occurred even though she agreed to seek out alternatives, such as additional clinical work on the cardiac critical-care service, and to pay for any additional costs that might be incurred. She eventually sued the university in 1993 and won her suit in 1995, collecting nearly six figures in monetary damages and requirements that the school develop alternatives to the animal labs for others who objected on ethical grounds.[6]

The professor shrugged his shoulders, said "How absurd," and tried to convert the discussion into a rant against lawsuits and lawyers. Many in the audience were happy to join in the fun of mocking lawyers, perhaps as a way to once again relieve the tension engendered by the difficulty of the ethical question before them.

My last effort was to note that the dogs on the lab tables did not just magically appear there. I argued that in order to conduct an honest ethical analysis, we need to be cognizant of the harms done and suffering experienced by the dogs through all the steps of the experiment process; we need to appreciate the costs in pain and distress paid only by them for this academic exercise. So I began to elaborate the details. Many of these dogs had spent some portion of their lives lost or abandoned on the streets, likely without access to regular food, water, or protected sleeping locations. Once captured, they were held in a strange and certainly stress-provoking shelter environment for several days and then trucked to the university, where they were housed in another strange environment for more days, this time in individual cages stacked two high. On the fateful day when their already unlucky lives were to be converted into an educational experience, they were carted to the lab, anesthetized into unconsciousness, instrumented with the devices that would transduce the relevant measures, dissected and studied, and then killed and disposed of. The animals most likely selected for this process, I said, were dogs that were friendly to humans, indicating their likely status as former pets. I asked the audience whether this characteristic made the process easier for them or expanded the harm. Was this an exploitation of vulnerability and a dog's expectation of care and support? Did it matter ethically?

No responses were offered, but the room became very still. Finally I suggested that the claim that the educational goal of this tragic sequence of events somehow neutralized pain and even added value to these otherwise "wasted lives" was a fantasy meant to deflect attention from the harm-producing realities of the dogs' laboratory journeys.

From this and many encounters like it, I learned that attempting to change the mind of my opponent or hoping to get some recognition of the validity of the ethical issue in question was futile. The primary value of these encounters was to provide the more open-minded members of the audience with a different set of facts and perspectives than they might otherwise be exposed to. However, as the years went on it became ever more difficult for researchers in these debates to simply make the blanket claim that animals are inappropriate objects of formal ethical concern. Increasingly, research from the areas of ethology, cognitive neuroscience, and comparative psychology confronted both the scientific establishment and the public with images of what looked to be altruistic dolphins, grieving elephants, intentionally deceptive monkeys, and laughing rodents.

As a consequence, arguments against animal protection initiatives shifted from denying animals' moral status to critiquing the regulatory barriers that had to be negotiated to get approval from oversight committees. It was frequently argued that it was more difficult to get an animal experiment approved than a human experiment. The nature of this irony was left implicit but clear: animals are insignificant morally as compared to humans and are not worthy of the time required for a detailed protocol approval.

At these times I attempted to interject the notion that animal protocol approval *should* be problematic at least, because our knowledge of the subjects' experience of the experimental process is far more uncertain than it is in most human experiments. After all, we are humans and our knowledge of what that is like to be one can in many cases serve as a source of protection when we're estimating the experience of harm to others like us.

In my role as director of the Research Ethics Service Project, I realized that hoping for many opportunities to speak directly about the animal question was a mistake. Many safer topics were in much greater demand. Thus I began including discussions of animal ethics in all my bioethics presentations. I did this by identifying the animal issue as one of a series of very important bioethical questions that all have to do with a fundamental question: what characteristics must an entity possess in order for it to gain access to the moral protections provided by the norms of a society or subgroup? In that sense the animal question is inherently related to such issues as the definition of death, the ethics of abortion, and the appropriateness of research with human embryos and stem cells.

My adding the animal question to this group of tormented ethical issues surprised many listeners. This reaction demonstrated the extent to which animal ethical issues had been split off from bioethical matters relating to humans. Examining reasons for this division was an essential part of these talks. Was it because some believed that animals were locked into a set of instinctive, fixed action patterns constructed by natural selection and therefore could not take personal credit for the intelligent-appearing behaviors that made up their behavioral repertoires? Was it because animals by and large were not able to feel pain or even possess preferences to avoid it? Was it because they were not able to anticipate the onset of harm and therefore suffer? Was it because they were not aware of death nor the fear of the closing of all future possibilities that came with that knowledge, so that the loss of their lives was an ethically irrelevant event? Was it because in their natural environments animals lived "with" nature and were not obviously involved in striving to improve their life situations? Or was it because animals were incapable of modifying their own behavior in response to a cognitively created moral code?

Taking this approach allowed me then to describe various ethical theories and their required capabilities for entry into the realm of moral considerability. For a utilitarian like Peter Singer or Raymond Frey, an organism's ability to experience pain or pleasure constituted the necessary capability. For a Kant-influenced theorist like Tom Regan, being the "subject of a life" was the standard. Demanding an even more complex array of capabilities—such as self-consciousness, reflective thought, or even membership in a moral community—would have the effect of severely limiting the kinds of animals (and humans for that matter) who were to be treated in strict accordance with high-level protective moral norms. Ultimately I hoped to stimulate people to ask, "Why is it the case that the more similar to a normal human being an entity is, the more ethical protection it deserves? Why are we the single standard? Why doesn't the need for protection play a significant part in the moral scheme?"

However, no matter the context of the presentation, the threat that I believed hung over my comments was the recognition that unless I was careful to express obvious respect for the pro-research arguments, whether they were deserved or not, I ran the risk of being identified as an animal rights advocate and having many listeners rapidly dismiss my credibility. To assert that animal rights could be a coherent ethical position was different from debating disagreements on other biomedical issues. In my experience, animal researchers' disdain for the concept of animal rights exceeded the level of derision evolutionary biologists expressed at proponents of intelligent design. I am quite convinced it was because I openly expressed doubt about the questionable validity of many of the animal models of psychological disorders that my term on the medical campus IACUC was not renewed. Although skepticism is the heart of scientific thinking, expressing it toward certain sacred methodologies could be the kiss of death for an IACUC member.

The presentations I gave to public groups as a part of the RESP initiative were less contentious than those on campus, although they also had less predictable outcomes. I had just completed a thirty-minute lunchtime lecture on animal welfare issues to the Kiwanis when a sturdy middle-aged man approached me. The lecture I had just completed was practical, emphasizing the nature of the developing protections for research animals that were becoming part of federal regulations. I had made no attempt to review the details of the various theoretical controversies. Instead my goal was to show that the changing public morality about the level of protection given to animals was slowly influencing research community norms.

The man approaching extended his hand and thanked me for coming. I could feel strength in his firm grip. "I grew up in North Dakota," he said. "I guess you could have called me a great white hunter in those days. I wanted to be a trapper. That was until one winter morning while I was checking my trap line and found a pheasant dead and frozen in one of the traps. My partner called it a trash animal and tossed it aside." He stopped for a second, looked up at me, and said, "But the bird had half eaten its leg off trying to get away. When I saw that, I knew that I couldn't trap anymore."

He wasn't finished. We sat down again, facing each other across the lunch table with its half-filled ice tea glasses and remnants of sandwiches. The room was empty except for the restaurant staff. "A few years ago I was hunting in northern New Mexico, where I shot the biggest buck deer that I ever saw. I watched him drop, but I was unsure that I had killed him. When I walked the one hundred yards to where he lay, I discovered that I was right; he wasn't dead. I had shot the deer behind its shoulder, shattering its spine. It lay there paralyzed." Again he stopped speaking. He looked at his hands, unfolded in front of him, then at me again. "When I walked up to the deer, he stared at me eye to eye. I saw the question in his eyes: 'Why?' I couldn't answer. I can't answer now. That look still haunts me today, right now as we speak. I know I had the legal right to shoot him, but at the same time it was deeply wrong to end his life."

After a few moments of silence he took a deep breath and stood up; we shook hands again. I waited until he left the restaurant before

I followed. I didn't want to meet him in the street. I didn't want any required social nicety to interrupt what we had felt together.

It seemed that he was trying to find a way to express his sadness for killing a deer that had grown large, robust, and strong through its own adaptive cunning and good luck. In his halting words he had described the conflict between a traditional social norm that encourages the killing of animals with traps and guns as a wholesome sport and the feelings engendered when for a few moments he allowed himself to see the struggle to live of a desperate pheasant and a paralyzed deer. My talk, while not obviously addressing the ethical issues around hunting, had facilitated his empathic engagement with the images of those animal encounters once again. He was able to reveal his sadness to me without fear of critique, suggesting that he saw me as a person who shared his conflict. He was of course correct. I was using these talks as a way to help work out my own quandary.

Another occasion involved a book club devoted to philosophical matters that met after work at a local laboratory. The group was made up of a few scientists and some nonresearch staff members. I was invited to present a lecture about the impact of the animal rights movement on scientific progress. It was quickly clear that many in the group were already certain that the rights movement was a philosophical frivolity doing great damage to scientific progress and the hopes of the medically infirmed. I took the position that while I believed animals had rights, the important question was which ones. I asked the group members whether they would agree that animals should at least have access to humane treatment by the researchers who had taken possession of their lives. There was ready agreement to that position. I then reviewed the requirements of the current ethical heart of the American and European animal research welfare systems, which are known as the "three Rs": use alternatives to animals if possible (replace); use as few animals as possible (reduce); and minimize the amount of suffering and pain experienced by the animals (refine).[7] Again, there seemed to be agreement. I asked if they thought applying the three Rs to animals in research was an act of charity and therefore optional, or whether they saw those requirements as the minimum standards demanded by professional obligations. I then asked whether it was coherent to say that animals in research have a "right" to the treatment outlined in the three Rs.

What followed was a rich discussion about philosophical rights and then a discussion of whether animals were always good models of human disease. The conversation continued among the attendees even as I bade the group goodbye.

But before I made it to the exit, a young man in his late thirties introduced himself as a biomedical scientist who worked at a privately funded laboratory and was a regular user of nonhuman primate and rodent subjects in his work. He did not want to discuss the rights issue but to tell me about a dream that he had been having repeatedly. I was caught totally off guard. In his dream, he said, he saw himself in a wide-open field where he felt very relaxed and calm. However, off in the distance he saw a human-size albino rat standing alone in the grass. He tried to approach the rat, which he described as absolutely beautiful, but it kept moving away from him. Finally he managed to touch it, and it dropped like a stone and was immediately dead. He recognized the obvious relationship to his work and the large number of animal deaths he caused. But instead of seeing this as a sign that he should reconsider some aspect of the work, he interpreted the dream as requiring that he work even harder at his research in hopes of justifying the many deaths and gaining absolution. That is, he felt that if he asked himself to suffer during the course of his research, it balanced the suffering he created in animals.

The extent to which the RESP was able to achieve the goals that I had optimistically set for it in 1994 varied according to the funding preferences of the differing research administrations that were responsible for it. Generally when there were any funding issues, RESP was the first program whose financial support was reduced. It was my clear sense that as long as the appropriate boxes could be checked verifying the existence of ethics training with the exact minimum hours of instruction required by federal regulations, the matter of research ethics, human or animal, was considered settled. Going further than that was definitely not a priority.

The RESP initiative received a substantial boost when in 1996 the philosopher Dr. Joan Gibson, an expert on end-of-life issues who headed the ethics program on the UNM medical campus, offered me an opportunity to collaborate with her on research ethics issues. Together we produced a successful conference on animal ethics in 1997. Soon after Joan's early retirement, Laura Roberts, MD, who had become director of the Health Sciences Institute for Ethics in 2000, asked me to formally join her institute. Dr. Roberts was a very well known and respected psychiatrist and human research ethicist. Her work on the ability of psychotic patients to consent to participate in clinical research was well known to me in part because Janet Brody, one of my PhD students, had worked with her on many of the initial studies. Laura's institute was housed in a beautiful converted home on the edge of the north campus and had its own staff of capable administrators and colleagues and a firm budget. Laura supported my work, encouraged me to expand my lectures and consultations at the Health Sciences Center, and supported my developing, with Barbara Orlans, a series of animal-research-focused conferences that were held at the University of California–Davis and Tufts University.

Unfortunately, Laura left UNM for the Medical College of Wisconsin and then Stanford University, and her replacement was a geriatrician interested almost exclusively in clinical bioethics matters such as advanced directives, not in animal research. As a result, I decided to retire from UNM and move my animal ethics activities to the Kennedy Institute of Ethics to continue working with Barbara Orlans and Tom Beauchamp. Tom supported my appointment as "institute affiliated faculty," which provided me an important group of colleagues.

Recently I presented an animal ethics lecture to a group of biology undergraduate students with declared interests in pursuing research careers. A week or so after the lecture, the professor who had invited me, Dr. Alexis Kaminsky, sent me a sample of the reaction papers she had asked the students to write. One student recounted an animal encounter she had as young girl. On a road trip with her family, they spotted a group of deer playing near the side of the road. She and her siblings implored their father to pull over so that they could get a better look.

As we approached them they all jumped over some very faint wire fence that we had not noticed; all except a smaller deer that could not jump high enough to get over the fence. I remember thinking that it was great that he could not get across because then we could probably touch it without it getting away. As we got closer the deer seemed more and more in distress, and my father ordered us back in the car. I was really upset that my father had ruined this opportunity for us. I will never know the real reason why my father made us leave the deer alone, but I believe that he felt bad for the little deer because the rest of the group had abandoned him and he hoped they would come back if we left. Sometimes I think I lack this sense—the one that reminds you to stop thinking about yourself and realize the distress you are possibly causing on others; I do not really think it matters whether the others are animals (including mice, rats, or birds!) or people.

This student was encouraged by her ethics class and my presentation to wonder why she was not sensitive to the plight of the animal, recognizing that something of ethical importance was missing. I could sense and empathize with the turmoil she was experiencing. This was exactly the kind of outcome I had worked to create with RESP. Small successes like this are what generate the feeling that all the work has been worthwhile.

∗   ∗   ∗

In addition to developing and then running the RESP after my return to New Mexico, I again served on two different Institutional Animal Care and Use Committees (IACUCs). As one of the few manifestations of concern for the welfare of laboratory animals at the university, the IACUCs were an obvious choice of venue for my efforts to make a difference in the lives of the animals used in research and teaching. The ethics training I had received at Georgetown increased my effectiveness as an IACUC member (and as chair, when I so served), but it also heightened my awareness of the inherent limitations of the IACUC system. While often frustrating, this awareness could be harnessed to examine the system's faults and failures, better understand their causes, and imagine and design corrective measures.

There is a widespread perception that the IACUC system put in place by the 1985 amendments to the Animal Welfare Act is more than adequate for ensuring that animals used in research are not subjected to unnecessary pain and suffering. As noted earlier, the law requires investigators using animals to reduce to a minimum the pain experienced by animals, to use the fewest number of animals necessary, and to consider the use of alternatives to animals if they are available—the set of requirements known as the three Rs. In addition, all research protocols must be approved by an institution's IACUC, and the committee is empowered to observe firsthand any laboratory and teaching procedures. Because this oversight structure seems so comprehensive, it is easy for researchers to believe—or simply to claim—that it fulfills the animal protection goals of the legislation. As a result, one commonly encounters views such as that of professor of neuroscience Craig Berridge, who is also chair of one of the animal research oversight committees at the University of Wisconsin Madison: "Animal research is a heavily regulated and overseen process and I think everyone who does animal research feels they're balancing the need for and desire to alleviate human suffering [with the need] to minimize animal suffering."[8]

However, many powerful factors at work in the lab—human ego and curiosity, the drive to advance science, the fear of human disease and suffering, strong academic mentors who model exaggerated confidence in the value of animal models, unquestioning acceptance of traditional methods—directly oppose efforts to expand the welfare of animals. These factors encourage investigators to ignore the ethos behind the law (that animals have value independent of their scientific utility and have rights of some kind), to resist complying with the regulations, and to exploit the weaknesses of the IACUC system.

As an IACUC committee member, I was repeatedly exposed to examples of these dynamics. While I retained my underlying faith in the potential of the IACUC system to encourage serious thinking about animal welfare and treatment, I came to see that this outcome is far from assured, as is often claimed, and is in fact highly contingent. Five issues related to IACUC oversight stand out as the most problematic, as arenas in which the promise of the system is often left unfulfilled and the proper approaches can make all the difference: the identification and categorization of pain and distress; what constitutes adequate housing; the role of the veterinarian; the handling of misconduct; and the search for alternatives to animal use.

As part of the effort to reduce the amount of pain that animals experience in research, the protocols submitted for consideration must describe the pain likely to be experienced by the animals to be used. Each animal is placed in one of four categories according to the pain it is expected to experience: no pain (category B), brief pain like a needle stick (category C), pain relieved with anesthesia or postprocedural analgesia (category D), or pain that will not be relieved for scientific reasons (category E). Category E designation requires further scientific justification before the protocol can be approved.

The problems with this rational-appearing system are fundamental. One problem is that this system focuses its attention only on the circumstances of the specific experimental interventions that are the primary focus of the research. These might include an animal's reaction to a toxin, the aversiveness of a loud noise in a startle-response test, or the distress produced when a cancerous tumor is implanted. It does not require that all the events that surround the acquisition, transport, and housing of an animal, which also involve harm, be categorized and entered into the cost–benefit analysis.

Further, the standards for judging pain are vague. The regulations hold that unless one has information to the contrary, the judgment standard of the degree of pain to which the animals are exposed should be what a human would experience in a similar situation. But the type of human to be used for the comparison is not specified. Should the comparison be with the reactivity seen in a human infant, a well child or adult, or an adult with a similar disease state as the animal model under study?

And then there is the intrinsic problem of knowing the subjective experiences of beings unable to communicate with humans. It was many years after the philosopher Thomas Nagel had published the classic paper "What Is It Like to Be a Bat?" in 1974 before I seriously entertained the judgment-of-pain problem and the relevance of this paper.[9] Nagel describes in a thought experiment the futile attempts a human might make to try to understand the world as perceived by a bat. What use would come of blindfolding ourselves, using a cane to approximate orientation by echolocation, attaching winglike structures to our arms, or hanging upside down during sleep? At best, Nagel points out, our attempts would tell us what is it like to be a human trying to perceive the world of a bat.

This general idea has become quite revolutionary in psychology, the philosophy of mind, and cognitive neuroscience as the concept of *embodied cognition*.[10] That is, the brain does not just sit isolated and separate from the rest of the body, processing information and directing action. Rather, the form and states of the body are active participants in cognitive activity. This means that accurately assessing the level and quality of pain perceived by an animal during an experiment would involve the highly complex and practically impossible tasks of detaching from one's own embodied cognition, entering the body-mind interactions of another being, and then somehow retrieving those experiences and interpreting them in one's anatomical and cognitive home.

Using a healthy human adult's experience as the measure of the pain and suffering an animal can be expected to experience doesn't solve the problem of knowing an animal's pain, because we really have no idea how close human pain is to the pain experienced by animals. Some have suggested that this procedure will overestimate the level, and others see it as providing a vast underestimate. An example of the overestimate argument is offered by the prominent philosopher of mind Daniel Dennett. Dennett contends that for a creature to suffer—that is, for its pain to be multiplied—it must not only be sentient but also be able to reflect on how the experience of pain will set back the pursuit of other important parts of its life and to be able to anticipate those setbacks.[11] Dennett doubts that there are many animals who have the cognitive complexity to fulfill these requirements. According to this argument, most animals will experience nothing but a series of unconnected pains over time that will not be magnified by anticipation.

Professor Bernard Rollin sees it quite differently. For him, an animal's possible lack of self-consciousness means it is locked into the moments of its existence. If these moments are occupied with significant pain, there is no way for the animal to escape the impact of the pain because it cannot, as a human might, shift its perspective away from the pain to a possible pain-free future. Thus animals do not *feel* their pain; they become their pain, their world is pain.[12]

These starkly different considerations call for nothing but humility when assigning a working level to the pain experienced by an animal. Ethically it would be better to potentially overestimate pain than to underestimate it.

Beyond the intrinsic difficulties of recognizing the presence of pain in another being, the regulatory system is essentially powerless to do anything about intentional underestimates of pain or underestimates resulting from systematic biases. Given that the regulatory system requires an investigator to further justify placing animals in any experimental conditions that involve unrelieved pain, most investigators understandably try to avoid a category E designation. This can result in what looks like a quiet conspiracy between IACUC members, who prefer to believe that animal research is a basically humane and pain-free enterprise, and researchers who share the same wish, as well as preferring to avoid having to further justify their work publicly. The administration of a research institution likewise prefers not to have to report animals used in category E experiments, perhaps fearing that the potentially publicly accessible statistic will attract the attention of protectionists, who may then ask to examine the projects further.

A protocol I recently reviewed from a world-renowned laboratory illustrates the problem. The research was about developing a nonhuman primate model of human affective disorders. The experimental methods described involved extravagantly harm-generating manipulations such as mother–infant separations immediately after birth, periods of socially deprived rearing, frequent anesthetic knockdowns, more than brief full-body immobilizations, deliberate and sustained provocation of fear reactions, introduction to stranger monkeys in social contexts that could involve various degrees of aggression and possible wounding, and so on. Despite the extent of the harm, the principal investigator placed the entire protocol in category C, which encompasses studies involving no more that momentary exposure to pain and distress, such as what occurs during an injection or brief blood draw. Putting this protocol in that category was like saying that the only significant harm for a human patient having her leg amputated was the postsurgical pain and not the cascade of difficulties and disabilities that would follow such a procedure.

Was the investigator intentionally being deceptive? I think not. He had become oblivious to the impact of submitting infant monkeys to such "rigorous" experimental treatments. Even more problematically, the IACUC agreed with that minor pain designation.

In another proposed study, researchers would purposely infect a mouse with a parasite in order to extract a fluid produced only by mature forms of the bug. The mouse would get sicker and sicker, and just before death the experimenters would euthanize it and collect the target fluid. This procedure was placed in pain category D, under the reasoning that the mouse's suffering was relieved in the end by euthanasia. In placing the study in this pain category, the experimenters apparently believed that the killing magically erased all the pain the mouse had experienced up to the point of death. In my mind, this kind of thinking is the height of indifference.

Housing is a particularly important issue in animal welfare, because once an animal is placed in laboratory confinement, few elements of its wild nature can be expressed, and even those few are rarely able to be integrated into patterns that give a life meaning. Discussing the proper way to house laboratory animals, the biologist and animal welfare advocate William Russell put it this way in 1956: "Captive animals usually know what is good for them, and our chief concern must usually be to provide them with the essential components of the environment from which we have removed them."[13]

Most centrally, this involves providing housing that allows for social interaction and significant movement. While attempts have been made to improve the housing environment where animals are held during experimental involvement, these attempts typically fall far short of providing many of the intricacies present in the more natural environment. This shortcoming is the result of many factors, including lack of knowledge about a particular species' natural environment and uncertainty about how to practically translate the most important dimensions of wild life into a confined laboratory environment. Researchers have also frequently expressed the concern that facilities that depart too far from older traditional caging methods will produce empirical outcomes that can't be compared with data collected earlier.

Housing is directly related to the requirement that researchers develop and implement plans to improve the psychological well-being of the animals in their charge. When this requirement was instituted for nonhuman primates in 1985, the reactions varied considerably and are instructive. While some researchers saw the benefit of improving the mental state of primates both for the monkeys themselves and for the validity of the data collected,[14] many others resisted, pointing out that creating new or modifying old holding facilities to allow for meaningful social interactions would result in the closing of many laboratories unable to meet the financial costs. Others argued that while the moves were well intentioned, they would surely result in injuries, some fatal, once highly aggressive monkeys were mixed together in groups.

Viktor Rheinhardt, once the attending veterinarian at the University of Wisconsin Primate Center, deflated many arguments for the status quo by showing that important psychological welfare improvements could be established with simple and inexpensive alterations. These included pairing compatible same-sex monkeys in connected individual cages and adding tree branches and plastic perches in order to increase usable living space. Instead of finding dangerous aggression, he found that slowly paced introductions of new monkeys resulted in few incompatibilities and enormous increases in prosocial behaviors like grooming. He made it clear that the resistance to change resided primarily within the individual researcher when he invoked the disarming cliché "When there is a will, there is a way."[15]

To put the issue of housing in proper perspective, we need to remember that even the most capacious housing is still captivity. Even though animals may be able to do things resembling wild behaviors while in captivity, the entire foundation for the lives for which evolution has prepared them has been removed. The circumstances of caged monkeys may be respectfully likened to those faced by the people of the Crow Nation after the tribe was forced onto the reservation and the buffalo disappeared. These changes left the tribe unable to organize around the buffalo hunt and wage territorial war with the Sioux and thereby develop the character of courage in their children. Chief Plenty Coup admitted that if a person unfamiliar with Crow culture looked around the reservation, he or she would see food being cooked, babies being born, children being raised, and homes being built and restored—but in terms of the Crow way of life, he said, "nothing happened."[16]

While all members of an IACUC have potentially important roles to play in balancing animal harm and the hoped-for benefits of a project, in my experience the person most central in driving the costs side of that analysis is the veterinarian. The veterinarian can explain with authority the consequences of experimental treatments and laboratory-related diseases and disabilities to English professors and public members. He or she can knowledgeably make judgments about whether a researcher is adequately trained to perform an invasive alteration of an animal's body and provide attentive postsurgical care. He or she is in a position to model to senior researchers and students alike the compassionate elements of a doctor–patient relationship. He or she can show how to identify and to some extent decode the evidence of pain and distress and what is necessary to alleviate it.

The veterinarian is also the person in the best position to judge a researcher's experimental character. He or she can identify the corner-cutter, the rough handler, the ignorer of student and caretaker worries—or as they are derisively known, "the cowboys." Conversely, he or she should know who the meticulous and caring scientist is, the one who actually supervises assistants and students with patience and knowledge, the one who is not afraid to use what have been called the six most sacred words of science: "I don't know; I need help." Providing this information to the chair of the IACUC—not as stigmatizing innuendo but as data grounded in observation—can help in the allocation of limited time and attention.

Unfortunately, some lab animal veterinarians can be negative forces instead of helping to elucidate the welfare impact of experimental procedures and discuss project value. From my years in graduate school to the present, I have personally worked with ten lab animal veterinarians. All had their virtues and limitations with respect to teaching students and senior investigators the skills necessary to recognize and treat medical conditions that actively interfered with their research projects. Some waited to be called upon to consult on a problem, while others believed in making their presence highly visible in and around laboratories, inserting themselves helpfully whenever possible even in the absence of a defined problem or specific invitation. Some limited themselves to keeping an approved project going, while others saw themselves as playing broader educational roles. These different approaches have major animal welfare consequences. The veterinarian who waits to be asked to consult will know less about the people involved and their work styles and instructional needs, and he or she will be less likely to be approached by an assistant wanting to report concerns. I have also seen veterinarians who protect the researcher from what they see as unnecessary IACUC member questions. They do this either by choosing to answer themselves the questions meant for the investigator or by discounting the value or appropriateness of others' questions.

Another important issue is how the veterinarian handles information from animal protection groups. If he or she adamantly rejects these groups as valid stakeholders, this can chill the contributions of members who have such allegiances but leave them unspoken. At one time I would have said that this problem affected mostly public members associated with the Humane Society (local or national) or an animal sanctuary. That is no longer the case, given the academic expansion of human animal studies programs, which are now at home in many different university departments.

It must also be said that veterinarians occupy a very difficult position in an academic research institution. First, while they often have academic titles and an association with an academic department, they rarely have a protected tenured position. Therefore they are vulnerable to conflicts with the research administration and institutional officials, whose own performances are centrally judged by the aggregate of research productivity their organization can generate and extramural grant support it can attract. Thus the veterinarian must tread lightly when making reports of problematic investigator behavior that might affect funding and publication numbers. I was once a consultant to an IACUC doing an investigation of a researcher who was charged with not having his protocol properly reviewed and approved before proceeding and may have even forged the approval signature of an official. In private, one of the university attorneys complained to me that he thought the research administrators seemed more interested in how to legally fire the complaining veterinarian than how to sanction the researcher if he was found to have violated proper procedure. How many such stories would an attending veterinarian have to hear before he or she either leaves a position or pulls back from fully reporting misconduct?

It also matters a great deal how IACUCs and their associated institutions handle researcher misconduct. In the years before protective laws and regulations were set in place, gross errors, carelessness, and outright animal mistreatment were shrouded from view unless a concerned person, by accident or intention, was able to view the workings of a scientist's laboratory or the conditions of an out-of-the-way holding facility. Today required unannounced inspections (when they are truly unannounced), coupled with a professional staff of technicians and veterinarians able to enter labs openly, have improved the ability to discover problems. Once a problem is discovered, the question becomes how to best handle the suspect situation. Obviously, misconduct comes in many different forms and should stimulate different degrees of concern. The investigator who uses a recently out-of-date drug and the one who doesn't adequately alert assistants about the need to shorten a food-deprivation period both fail in their responsibilities to their research and their public promise to protect research animals, but this failure is not egregious. In cases like these, a furrowed brow, a brief informative comment, and a reminder that it is the animals that pay the price for our inattention will often suffice for the well-intentioned investigator.

However, if a senior researcher is discovered to have carried out an experiment from beginning to end without IACUC approval, using the justification that "everyone does it," or fails to euthanize severely injured or sick animals as directed by the veterinarian in order to "get a little more data," or repeatedly directs an uncertified member of the research team to conduct surgery, different consequences are called for. These kinds of actions reflect contempt for the ethical consensus and public concern that led to the legal protections in the first place. The actions should be looked at as violations of research animals' right to be treated in accordance with the established rules and regulations. *Right* is the correct concept, because a researcher cannot independently decide he is going to override the requirements simply because he believes he has a good reason to do it. The law, in a sense speaking for the animal, trumps that possibility.

It is in these severe cases where I have seen institutions stumble in their responsibility. Too often administrators show an unwillingness to acknowledge openly egregious behavior and to consider suspending or terminating the person's animal privileges or perhaps even his or her employment. The dynamic that contributes to this mishandling is the all-too-common belief that if brought to the attention of the institutional community or general public, knowledge of the misconduct will lead to loss of the institution's reputation and result in collateral damage. Revelations can lead to questions about why the person was hired in the first place, as investigations often reveal a history of previous misconduct that was ignored during the hiring process. There also can be concern that the misconduct may fall under a state's animal cruelty law and lead to a highly visible prosecution. As a consequence, those responsible for institutions are tempted to go into a self-protection mode with the goal of keeping the incidents quiet, minimizing their seriousness, or possibly encouraging the person in question to leave before public discovery.

In the latter case, large financial incentives are provided and confidentiality agreements are signed, and the offender disappears from campus.

The R of replacement—the AWA's imperative for researchers to meaningfully explore the use of nonanimal alternatives—is in my experience not taken seriously enough. Even though we now have vast searchable information resources, few researchers take the time to perform even cursory searches of the relevant databases. I continue to be struck by how relatively few researchers are aware of the online information sources provided by such organizations as the Johns Hopkins Center for Alternatives to Animal Testing (CAAT), the UC-Davis Center for Animal Alternatives Information, the Animal Welfare Institute (AWI), the Fund for the Replacement of Animals in Medical Experiments (FRAME), and the Animal Welfare Information Center (AWIC). Despite their availability, much of the searching for nonanimal alternatives is limited submitting a couple of obvious search terms submitted to PubMed, without making use of Boolean term connections and varied multiple searches. Requiring researchers to indicate in their protocols the terms they used in their searches is a meaningless exercise unless the IACUC has access to expertise like that of a reference librarian who is capable of determining the adequacy of the method described. More fundamentally, the requirement is for the researcher to "consider" the use of alternatives. Given scientists' tendency to default to long-established and familiar methods, the inclination to use new alternatives is limited and "consider" is a weak ethical principle.

For it to make any practical difference, a researcher must conduct the search for alternative methods and subjects at the very beginning of the research process, before making a firm commitment to a particular experimental strategy or method. The basic problem is that such a state of experimental neutrality is unusual. Researchers have more than likely been trained to use a particular set of methods and have strong preferences driven by previous success. They might also have a previously established animal colony in their lab and a need to continue to support its existence. As a result, when new experiments are in the planning stage, those familiar methodologies—including particular animal models—have a clear choice advantage and are not likely to be displaced by unfamiliar alternative methods.

During the fifty years I have spent working in and around animal research, many serious and profound philosophical analyses of the ethics of animal use have been put forward, and a body of animal law and regulation has been developed that on its face intends to advance more humane treatment of animals in research. However, the extent to which these policies and ideas actually result in significant animal welfare protection remains contingent on such factors as circumstance, personalities, the member composition of IACUCs, and committee leadership. In my experience, a local Institutional Animal Care and Use Committee can provide little protection unless the following conditions exist:

1. Committee members value animal lives at least as much as they value animals' usefulness in research.
2. Committee members consult more than the limited information provided by the research team when they prepare for a meeting, and they possess the resources to familiarize themselves with the research area and its ethical norms.
3. The committee meeting environment encourages members to voice questions, even those that may seem naive. (Nothing chills discussion more than rolled eyes and condescending dismissal.)
4. Committee members actively resist the tendency to defer to highly placed professionals or university colleagues. If deference to authority appears to be a problem, the committee is open to the possibility of using secret ballots in the decision-making process.
5. The committee has ready access to outside experts, including ethics consultants, who can evaluate protocols that are problematic or potentially sullied by conflicts of interest.
6. Research protocols include a thorough evaluation of the potential for harm at all significant contact points, including acquisition, transport, housing, experimental manipulations, postmanipulation conditions, and final disposition.

7. Committee members have access to information about the quality and impact of an investigator's previous or related research contributions.

8. The committee conducts postapproval review through cordial but truly surprise visits to labs and field sites and evaluates whether the methods observed actually comply with what has been approved by the IACUC.

9. The committee receives regular continuing education that includes relevant ethical theory and its practical application and not just animal law and regulations. Improving knowledge about methods of pain identification and control should also be high on the list of educational priorities.

10. When a committee member resigns or rejects reappointment, a university official follows up with an exit interview and remains open to hearing heartfelt criticism of committee process and decisions.

11. Accusations of serious wrongdoing or misconduct are publicly investigated by the IACUC and central administration, and appropriate sanctions are imposed if culpability is proved.

As should be clear, being a member of an IACUC is unlike participating on most other university or institutional committees. Membership requires a willingness to regularly go beyond the typical perfunctory review of provided information. It requires asking what some might see as annoyingly simple or intrusive questions and acting against tendencies of deference and reflexive trust. But mostly, in my opinion, it requires taking the position that harming animals for science is prima facie wrong but being open to being convinced by the evidence in a particular case that the research should proceed.

*   *   *

My decision to move away from animal research in the 1990s and to devote myself to the complexities of animal research ethics has always seemed like the right one, even though it had its costs. Although I always tried to present myself as an educator and reasonable critic of poorly designed and executed animal research and not a frank animal-rights advocate, the line of detractors and enemies grew. Bernard Rollin of Colorado State University warned me about this process when I first discussed my ethical concerns about animal research with him in 1992. He pointed out that while a philosopher like himself was expected to challenge carelessly applied ethical justification, a researcher who "changed sides" like me was likely to be seen as a dangerous turncoat. No doubt he was correct. But I have no regrets about my choice.

8

Reformation

*Even a purely moral act that has no hope of any immediate and visible political
effect can gradually and indirectly, over time, gain in political significance.*

VACLAV HAVEL

Compared to the state of affairs that existed when I began graduate
school in the late 1960s, animals used in research and teaching are
now afforded far more regulatory protection intended to reduce the
harm and distress they may experience. This change is a result of
both the animal welfare oversight system first put in place in 1985 and
discernible shifts in attitudes about animal use among members of
the public in general and some scientists. However, the slow evolution in values and views that has occurred in the last fifty years has
yet to tip the balance in favor of the animals. Most scientists—even
those who purport to have the best interests of animals in mind—
still hold fast to the presumption that animal use is justified a priori
because it yields considerable benefits for humankind. The harms to
animals that flow from this stance, abetted by career ambition and
willful blindness to the actual welfare consequences of animal acquisition, captivity, and experimentation, remain considerable. And,
as the previous chapter shows, the IACUC oversight system is severely constrained in its ability to limit the nature and extent of these
harms. Despite lab inspections, the availability of veterinarians, protocol review by committees including nonscientists and members
of the public, and the three Rs' guidance, animals still experience

intense and unrelieved pain, monkeys are still raised intentionally in devastating social deprivation, and animal lives are used up often with little to show for their sacrifice. All this takes place well within the acceptable legal and regulatory boundaries of the AWA and the IACUC system.

What else can be done? Despite my long-term involvement in an oversight system that has been only modestly successful in achieving its aims—and may be seen as standing in the way of more fundamental change because it promotes the impression that what's being done is sufficient—I remain unreservedly optimistic about the possibility that science, and society as a whole, will come to take seriously the notions that animals are not just property, that they have rights of some kind, and that appropriating animal lives for human use should always elicit ethical analysis that leans toward abstinence as the starting point. There are many signs that the ground really is shifting, particularly now with seemingly continual revelations of animals' surprisingly sophisticated cognitive abilities and rich, often humanlike emotional repertoires and social interactions. This new knowledge makes it possible to better elaborate and make visible the harms caused in the service of research and ensure a more realistic cost–benefit analysis. In what will likely become a pivotal moment in recognition of the costs to animals, a prominent group of neuroscientists meeting at Cambridge University issued what has come to be known as the Cambridge Declaration, which stated, "The weight of the evidence indicates that humans are not unique in possessing the neural substrates that generate consciousness. Nonhuman animals, including all mammals and birds, and many other creatures, including octopuses, also possess these neurological substrates."[1]

Further, on the benefit side of the decisional equation, mainstream science is tending to be more skeptical about the value of animal research. I am now seeing more than a few critiques reporting that much preclinical animal data fails to effectively translate to human biomedical use. In a 2014 article in the *Institute for Laboratory Animal Research Journal*, for example, Joeseph Garner asks, "Why do over 90% of behavioral neuroscience results fail to translate to humans, and what can we do to fix it?"[2] Elias Zerhouni, former director of the

National Institutes of Health, puts it more bluntly: "The problem is that it [animal research] hasn't worked." He implores researchers to refocus their research efforts on studying human disease in humans.[3] These and other publications call out for the return of skepticism in how we choose to pursue our animal work and interpret its significance. There is nothing more fundamental to the scientific enterprise than this.

It may come as no surprise that I see ethics education and training as a key to making progress in the way we treat animals. Unfortunately, laboratory researchers, like most health- and science-related professionals, live with the moral provocation provided by the closed door. For the most part we do our work out of sight of others but in the full view of ourselves. In that chamber we can possibly find the best of ourselves as we negotiate the forces created by frustration, whim, fear, desire, conformist pressure, and competiveness. Or the solitude can provide protection from the way these forces can disfigure our concern for unjustified harming of the other. What often makes the difference is the continual development of an informed moral identity; when we do not seek to grow as moral beings, we have failed our professional responsibility and are as vulnerable to those forces as the animals are to our choices.

The problem is that developing an informed moral identity is far down on the list of priorities for most people training for and entering scientific and professional careers. For too many, ethics training remains an optional adjunct or a tangential box-checking requirement. Without the encouragement to see themselves in moral terms and subject their decisions to serious ethical reflection, it is far too easy for young scientists to simply absorb and internalize the established herd consensus, the impoverished ethical framework that surrounds them at every step of their education and training. As part of a talk titled "Putting the Ph Back in the PhD Degree," I have in the past described the damaging retreat from philosophical and ethical sophistication in graduate research education. I found this message to be generally welcomed by many students, who indicated that they were somewhat aware of the deficits of their professional preparation but were in a quandary about what to do about it.

Early in my professional life I moved rapidly away from a reason-ably settled set of ethical beliefs and cautions to uncritically take up the norms that I saw around me. Where animals in research were concerned, I set aside ethical starting points such as that it is prima facie wrong to harm beings unable to protect themselves and to use professional status and raw physical strength to overcome their objections to being manipulated. I put aside the belief that we must show meaningful gratitude to those who have done us a service at great cost to themselves. I ignored the sense that I was accruing a debt that would eventually press for payment. Those responsibilities were replaced with all-too-well-practiced glib statements of the value of what I was accomplishing. Establishing and maintaining "the priv-ilege of distance" was also essential. I learned to avoid direct contact with the animals being studied in order to avoid facing the difficult evidence of what was being done in the name of science and at my direction. It took many years of experience and study to recognize that I had lost my moral bearings, to learn to see the animals as ani-mals, and to reflect on how basic principles of the common morality could be applied to my life and work. I have often wondered how it would have been different had I been exposed early on to a diversity of moral models or explicit ethical training.

Nevertheless, more thoroughgoing ethics training for scientists and a resurgent concern for ethical self-questioning among all citi-zens, as important as they are, will not be sufficient to curb the mis-treatment of animals, especially if we consider the broader context of human–animal relationships, which includes the food system's industrialization of meat, egg, and milk production, our entertain-ment preferences, and the thoughtless destruction of what remains as habitat for animals living their own natural lives. We will need to rethink and re-form the entire framework of beliefs that underlies our relationships with nonhuman species.

Central among these beliefs is the assumption that humans are not only at the top of the "evolutionary ladder" but also uniquely different from all other living things in morally relevant ways. It is assumed that from this special status come privileges that are ours alone; among these is the right to use other species for purposes that go far beyond nutrition and survival. This belief has come to be so taken for granted that when it is made explicit, as in the following statement from a 2014 article in the *Journal of Primatology*, it almost comes as a surprise: "As long as we believe that a human life is more valuable than [that of] a fish, fly, mouse, or primate, some experiments will be performed on animals before exposing humans to risk."[4] Here, in condensed form, is the basis for that faux-ethical notion that has remained completely intact in primate and biomedical research since well before I began my research career: if a research project is unethical to do with humans and has the potential for being scientifically useful, it is automatically ethically acceptable to conduct it with animals. As long as humans continue to believe in our special status and unique membership in the moral community, research on animals will continue to be performed with this justification, and the harms that follow from the associated discounting of the moral significance of animals' capacity for suffering will continue.

Other models for perceiving the place of humans relative to other species exist. I have had the privilege of being exposed to one such model, in which humans are taken to be members of the web of life, equal in status to the other members. This ethos acknowledges humans' special abilities, but rather than conferring on us unique rights, these abilities give us profound responsibilities.

I met Robert Lewis, governor of Zuni Pueblo, when he gave a lecture at an Earth Day celebration on the university campus and later at his official office on the reservation. The Zuni believe they have occupied their land in western New Mexico for thousands of years and that they emerged into this world from the Grand Canyon. At Lewis's office we discussed the proper relationship between humans, animals, and their earth mother. One of his first comments was that he thought many non-Indians "looked at their feet too much." Since they are interested primarily in their own place on the earth, said the governor, they focus on the footprints that are indicative of their path. "We must also look out from ourselves into the sky, mountains, deserts, and water," he said. "There we will see that we are only a small part of the world's beings. We are not the peak, we are a part."

In this context I described to him my quandary as an experimentalist. Did I have ethical authority to take the lives of animals for my work given how hard it was on them and the only slight chance that it would actually benefit anyone? He did not answer my question directly; in the Zuni way he told me a story instead. When he was a young boy, his father asked him to hunt piñon nuts with him before the snow came to the high desert forests. The nut is bean-shaped and about the size of a small raisin. After careful roasting, its thin shell can be pierced to access the soft center. The nut is very important to the Zuni and other southwest tribes as a food source, particularly for winter sustenance. It had been a dry growing season, and the nuts would be scarce. So Robert knew this was an important invitation.

Young Robert watched his father packing small sacks of grain and tying them to the backs of their horses. He began to wonder how long they would be gone if they had to take extra food for the horses. As they rode, his father told him that because the crop was small, they should keep a lookout for packrat nests, because the rat also collects the nut for the upcoming winter. He told him to watch for piles of debris stacked close to a fallen and decaying log. It wasn't long before Robert spotted a nest, jumped off his horse, and ran full tilt toward the mound; he would now do his duty for the tribe. He fell to his knees and started to claw furiously through the pile with his hands in order to reach the chambers full of nuts his father had told him were there. It was only seconds before he shouted in pain as he encountered the cactus spines that were scattered throughout the pile.

As he cried, his father quietly arrived and began to gently remove the spines. When his tears began to subside, his father reminded him that the mound was the packrat's home and that he protects it from intrusion with the cactus spines. Once his eyes were dry, his father showed him how to carefully move parts of the debris aside to expose large subterranean chambers that were indeed full of piñon nuts. After he filled several small burlap bags, his father blocked his reach and told him to stop, saying that they had taken enough. Robert asked why, since there were so many nuts left. Again, his father reminded him that the rat had collected the nuts so it too could survive the winter, just like them.

As he sat wondering, his father returned to the horses, loosened one of the grain sacks, and returned with it to the nest. As he watched in amazement, his father slowly refilled with grain the chambers they had just emptied of nuts. At this point his father intoned the fundamental lesson: "One cannot just take from animals and nature, one must also give back."

Having finished his tale, Governor Lewis looked at me and said simply, "Reciprocity."

This notion of reciprocity was of course not born with the Zuni, nor does it exist only in the Native American tradition. In Western ethics the idea is perhaps best displayed within the concept of the "moral residual" or "moral trace." Here the idea is that even if an agent weighs and balances the potential benefits and harms of an action, finds that the scales fall largely in favor of benefit, and then proceeds, the ethical obligations relevant to that act have not been completely absolved. Because a harm has been foreseen and knowingly created, a residual responsibility to ameliorate that harm remains regardless of the extent of benefits produced. For example, if a stranded and dangerously dehydrated traveler breaks into an uninhabited cabin to access a water supply that is visible through the window, the trespass is surely a permissible one. Still, a harm has been done to the owner of the cabin that requires attention. Some philosophers consider that all that is required is a simple apology, while others believe more extensive redress is required.[5] This example assumes that both the endangered traveler and the owner of the cabin have moral value and the loss to the owner has moral meaning and significance. With the case of animals, only a small percent of experimental use involves study of an actual imminent danger to humans, and the harms to the animals commonly involve significant setbacks in their welfare, including ending life itself. It seems to me that this situation creates an enormous residual debt that calls to us for discharge.

In Governor Lewis's example, the moral residual was the vulnerability that remained for the rat community if the stores of nuts, depleted by his and his father's collection, were now inadequate for winter survival. Instead of simply walking away from a reduced nest and rationalizing that humans needed the sustenance more than the rats because of their higher status, the father tried to minimize the residual harm. Replacing the nuts with edible grain was the payback attempt.

I asked the governor why that obligation did not also apply to other animals that might raid the rat's nest. He responded that our added capabilities create a greater burden of responsibility to live out the principle of reciprocity.

How shall we take on the burden of residual debt in the context of animal research? Many of my colleagues express, as I did for many years, the belief that they have discharged all their ethical responsibilities to animal subjects if they at least attempt to follow the ethos of the three Rs by seriously selecting the experiments that they conduct and making some tangible effort to seek less distressing alternatives. I have also heard others publicly acknowledge that doing research on animals is a "privilege" and through this utterance make a promise to promote respectful engagement. Others point to the future possibility that their research findings might find their way into veterinary medicine and ease animal pain and disease. Wildlife biologists who purposely disturb or take the lives of free-living animals for study express the hope that the information gained will ultimately foster conservation and the continued survival of a species. All of these options move in the direction of recognizing that something unique and ethically weighty occurs when we choose to manipulate the lives of animals in the context of research. However, agreeing to be influenced by the three Rs can become a minimalist exercise; acknowledging the privilege of doing research is of little use unless we reflect on the moral risk created when we see ourselves as capable of claiming such a privilege independent of animal interests; and the development of future veterinary or conservation benefits from our work is of course possible, but typically much more unlikely than we would like to admit. Animal research needs an ethical revolution if it is to live out a principle like reciprocity. Vague good intentions, promises, and furtive hopes may create an air of aspiration but little relief for the animals.

The story of the ethical evolution of human research shows us the needed direction. It wasn't until Henry Beecher's 1966 paper[6] that paraded the starkly unethical behavior of researchers, who in their minds were surely devoted primarily to human welfare, that things begin to change. Beecher's twenty-two dark experimental revelations provided the early momentum, and the revelation of the Tuskegee syphilis studies the final impetus, for the creation in 1974 of the National Commission for the Protection of Human Subjects of Biomedical and Behavioral Research. On September 30, 1978, this committee released the Belmont Report,[7] which laid out the foundational ethical principles upon which federally supported human research in the United States was to be based. The principles of respect for persons, beneficence, and justice encapsulated the essential moral requirements of protecting the autonomy and dignity of human persons, balancing risks and benefits when planning research, and ensuring that the costs and benefits of research were distributed justly throughout the population. Though not without criticism, the Belmont Report successfully transferred the duties of protection from individual researchers, with their inconsistent sensitivities, to the shared guidance of a basic set of ethical principles already embedded in the public morality. That type of foundation is essentially absent from animal research. There has been no move of which I am aware to create a federally appointed national commission to study animal research and produce a set of guiding principles like those in the Belmont Report.

A sign that such a move is no longer beyond the realm of possibility, however, emerged in 2011. In December of that year a committee of the Institute of Medicine issued a report,[8] requested by the US Congress, stating that chimpanzees were mostly no longer needed in biomedical research and ought not to be used in research unless it could be clearly shown that their use was absolutely necessary to advance knowledge about life-threatening conditions. Before using chimpanzees, researchers had to show—not just assert—that there was no way to ethically conduct the work in humans or nonape models. If chimpanzee use was in the end found to be necessary, the animals had to be housed in environments that met standards based on the chimpanzee natural environment (i.e., the housing had to be "ethologically appropriate") and not researcher convenience. Finally, if the necessary studies were behavioral in nature, the chimpanzees had to provide evidence of voluntary acceptance (i.e., acquiescence) before participation. The committee also took the opportunity to state unambiguously that trying to assess the value of animal experiments is not just an exercise in biology but must be also treated as an ethical issue that concerns itself with animal welfare costs. In other words, science and ethics cannot be separated. Francis Collins, the director of the NIH and a prominent scientist in his own right, accepted the report within two hours of its release.

Developing a Belmont Report for animals need not be a futuristic fantasy. If professional organizations and individual scientists demanded this kind of crucial clarification, it would happen. The Institute of Medicine report showed that informed scientists provided with sophisticated ethical guidance are completely capable of putting aside self-interest and traditional experimental assumptions and working to expand the web of ethical concern. While we cannot give back to the animals the lives that we take nor fully relieve the suffering we inflict, we can take a giant step toward reciprocity by demonstrating sincere regard for animals' being and rights.

Epilogue

As I noted in the preface, I searched for the sentence with which to begin this book shortly after euthanizing Donna, one of the surviving members of my stump-tailed monkey social group. As fate would have it, the two remaining members of that group, Millie and Lala, had to be euthanized as I was completing the manuscript. These monkeys were a mother-and-daughter pair who by that time had shocking white hair and moved around somewhat more slowly than they had in their prime. They had remained warmly involved with one another, mother Millie showing only an occasional fit of temper when her daughter displaced her from a favored sitting location or performed some other slight invisible to me. They had continued to delight in the array of fruits and vegetables that were offered to them buffet-style twice each day; Millie, a feral-born monkey, preferred green vegetables like spinach, snow peas, and green beans, while Lala opted for sweet fruits like oranges and strawberries.

The first sign that something was wrong came when Lala suddenly started to behave oddly. She began to walk very slowly while obsessively consuming small bits of food that she found scattered on the hay-covered floor. She did this slow food-wandering for hours at a time, passing up typical periods of sleep, mutual grooming, and sunbathing. She made no guarding movements that we could recognize as signs of acute pain, yet she appeared very distressed. The veterinarian had no clinical clue about what was going on and was concerned that aggressive clinical intervention would just cause more distress. I think this judgment was grounded in both sound clinical reasoning and a natural reluctance to handle a monkey known to be infected with herpes virus B. He recommended only watchful waiting.

But during the next twenty-four hours, the syndrome—or whatever it was—worsened. Lala began to give indications of having some abdominal pain, so some blood was drawn for testing. The results gave evidence of broad inflammation and multiple organ failure. We were both reluctant to stress an elderly monkey who had a limited remaining life span by initiating invasive treatment, so we decided to proceed with euthanasia.

Once we agreed on this, the discussion turned to Millie. Although nearly forty years old, she was apparently physically well. She showed some indicators of mental decline, appearing briefly confused and disoriented at times, and walked with a noticeable stiffness, especially on cold mornings. She warmed her body in the sun for many hours each day, and after the sun went down she stayed near the heating vent in the block house. Millie had been purchased from an importer who had captured her in South Asia and shipped her to New Mexico, where she immediately entered the monkey group I was forming. She had had only one offspring, Lala, and was an integral member of the group as it developed and added new members. She was not a dominant female by temperament but was a preferred grooming partner of the individuals who were. She had never lived alone except for that arduous trip from Asia to Albuquerque.

Was it fair, decent, and humane to leave her to live out her remaining life alone, with only human contact? Finding another monkey that this old girl could adjust to seemed improbable. Even if possible, it would start the problem again once Millie passed. Millie and Lala had depended on only one another exclusively for nearly three years, following the deaths of the other group members. Illustrating their connection, Millie would often let out a single high-pitched screech when she found herself sitting alone on a perch in the outdoor pen, and within two or three minutes Lala would emerge from the block house and climb up to the perch to join her. Lala would move in close and either groom Millie or lie down next to her, threading her arm around Millie's chest. After making some minor positional adjustments so as to maximize their bodily contact, the two would peacefully go to sleep together.

I turned to Gilbert Borunda, one of the two people who had cared for the group for well over twenty years, and asked what he thought. He agreed that for Millie to live alone without Lala seemed empty and cruel. I had talked a day earlier with Ector Estrada, the other long-term care provider, now retired, and he very reluctantly said that if we had to euthanize Lala, he thought Millie should be spared living alone without her daughter. The perspective of these two men was crucial. More than anyone else, they knew these monkeys. Because they had had to interact with them every day—feeding them, cleaning their enclosure, performing health checks and treatments—they had to develop ways of cooperating with them. They knew that using force and intimidation was not appropriate and would in the end make their jobs harder. Instead they needed to develop real working relationships, and they did. In so doing they came to know about the monkeys' preferences, annoyances, habits, and fears in a very personal way. So when they expressed the belief that the involvement and dependence that existed between these two was so intense that the loss of one would be devastating to the other, I knew to trust it.

Gilbert and I felt an urgent need to proceed because we did not want our ethical judgments to be overridden by the sense of anticipated loss that was beginning to well up along with the tears in our eyes. I did not want my personal preference for Lala and Millie's continued living to push me to wait a few more days in hopes of a sudden turnaround in Lala's condition.

As Gilbert prepared to get Lala and bring her to the treatment room, I noticed that he did not put on the thick leather catching gloves recommended for the process. He said there was no need, as he and Lala had an understanding. He wouldn't hold her too tight, and she wouldn't bite him or struggle to escape. As he opened the door to the block house he called her name, and she halted her slow-motion wandering. With his hands covered only with thin rubber examination gloves, he reached down, took and held her around her thick biceps, and picked her up slowly and carefully. He was right; she did not try to escape. I watched Millie as she reclined on a high perch, intently focused on the activities surrounding the capture of her daughter. She made no move to object.

Gilbert placed Lala in a sitting position on the examination table while supporting her back and loosening his grip on her arms. Lala sat quietly, looking like a chubby old lady. She seemed calm, looking up and down, left and right, and smacking her lips, apparently inspecting the people who surrounded her. Except for the vet, who had been here for three years, she had known all of us for decades. The vet injected the restraint drug ketamine; while part of accepted protocol, this step seemed unnecessary to me, because Lala was not going to struggle. Her calmness amid her distress was a clear expression of trust, just short of telling us in words "I know you won't hurt me." In response to the drug, her eyes became more active, with a repeating slight horizontal side-to-side movement. Her body remained still. After a few minutes, the vet administered the euthanasia drug through the saphenous vein on the calf of her left leg. She slid easily into death, making no sounds or movements.

During this process Millie had climbed down from her elevated position and came to sit in the hay close to where Gilbert had picked up her daughter. I told Gilbert that I wanted to take care of Millie. I entered the room wearing only rubber gloves—but I carried a catching net because I knew that she could be feisty at times. Gilbert assured me that I wouldn't need it. Doing his duty, the vet reminded me that she too was positive for herpes virus B and I should be more fully protected. I didn't acknowledge his caution. I didn't want there to be too many layers between Millie and me at this moment. Actually, I thought at the time that if she did bite me and I became infected and experienced serious medical morbidity, it would be just. Early in my research career I had supported a system that pulled her out of her home in Asia so that I could pursue a scientific passion without first seriously deliberating on the ethical costs to her and the others.

She sat quietly as I executed the same arm hold that Gilbert had used with Lala. Like Lala, she responded with calm and trust. The process proceeded as with Lala, and I watched as her life ended. Her eyelids closed halfway, and her lower lip went slightly slack to reveal the now bluish color of her gums. There was no way I could say to her, or to the others now passed, that I was sorry. It would have to do to leave this simple message: we must work hard to shed the methodological traditions and sense of unquestioned human privilege that allow lives like Millie's and Lala's to be taken over without deep ethical trial.

In response to a question about the use of animals in nonessential research, Charles Darwin wrote to Ray Rankester: "It is a subject which makes me sick with horror, so I will not say another word about it, else I should not sleep tonight."[1] I will stop here likewise, but only to catch my breath. The debt we have incurred has yet to be paid.

Notes

Introduction

1. K. L. Tyler and R. Malessa, "The Goltz-Ferrier Debates and the Triumph of Cerebral Localizationist Theory," *Neurology* 55, no. 7 (2000): 1015–24.

2. J. D. Hardy et al., "Heart Transplantation in Man: Developmental Studies and the Report of a Case," *Journal of the American Medical Association* 188, no. 13 (1964): 1132–40.

3. Augustine of Hippo, *The Confessions of St. Augustine*, trans. R. Warner (New York: Penguin, 1963).

4. R. Shattuck, *Forbidden Knowledge: From Prometheus to Pornography* (New York: Harcourt Brace: 1966).

5. Henry Beecher, "Ethics and Clinical Research," *New England Journal of Medicine* 274 (1966): 1354–60.

6. Now named the Wisconsin National Primate Research Center.

7. M. L. Stephens, *Maternal Deprivation: Experiments in Psychology—A Critique of Animal Models*, report prepared for the American, the New England and National Anti-Vivisection Societies, 1986.

Chapter 1

1. K. Kincaid, *A Walden Two Experiment: The First Five Years of the Twin Oaks Community* (New York: William Morrow, 1974), 5.

2. J. A. Osmundsen, "'Matador' with a Radio Stops Wired Bull," *New York Times*, May 17, 1965.

3. J. B. Watson, *Behaviorism* (Chicago: University of Chicago Press, 1930), 11.

4. Frank A. Beach, "The Snark Was a Boojum," *American Psychologist* 5, no. 4 (1950): 115–24.

5. H. F. Harlow and J. A. Bromer, "A Test-Apparatus for Monkeys," *Psychological Record* 2 (1938): 434–36.

6. H. F. Harlow, "The Formation of Learning Sets," *Psychological Review* 56 (1949): 51–62.

7. See S. D. Singh, "Urban and Rural Monkeys," *Scientific American* 221, no. 1 (1969): 108–15.

8. B. F. Skinner, "Psychology in the Year 2000," *Journal of the Experimental Analysis of Behavior* 81, no. 2 (2004): 207–13.

9. M. Kaplan et al., "A Restraining Device for Psychophysiological Experimentation with Dogs," *Journal of the Experimental Analysis of Behavior* 5, no. 2 (1962): 209–11.

Chapter 2

1. See D. Maraniss, *They Marched into Sunlight: War and Peace in Viet Nam and America, October 1967* (New York: Simon and Schuster, 2003).

2. G. C. Ruppenthal, et al., "Development of Peer Interactions of Monkeys Reared in a Nuclear-Family Environment," *Child Development* 45, no. 3 (1974): 670–82.

3. R. Melzack and T. R. Scott, "The Effects of Early Experience on the Response to Pain," *Journal of Comparative and Physiological Psychology* 50 (1957): 155–61.

4. H. F. Harlow, G. L. Rowland, and G. A. Griffin, "The Effect of Total Deprivation on the Development of Monkey Behavior," *Psychiatric Research Report* 19 (1964): 116–35.

5. H. F. Harlow, R. O. Dodsworth, and M. K. Harlow, "Total Social Isolation in Monkeys," *Proceedings of the National Academies of Science* 54, no. 1 (1965): 90–97.

6. H. F. Harlow, J. P. Gluck, and S. J. Suomi, "Generalization of Behavioral Data between Nonhuman and Human Animals," *American Psychologist* 27, no. 8 (1972): 709–16.

7. H. F. Harlow, G. L. Rowland, and G. A. Griffin, "The Effect of Total Deprivation on the Development of Monkey Behavior," *Psychiatric Research Report* 19 (1964): 116–35.

8. R. A. Gardner, and B. T. Gardner, "Teaching Sign Language to a Chimpanzee," *Science* 165 (1969): 664–72.

9. A recent article claims that much of what is understood about neuroscience is not valid because too few subjects have been studied in experiments. K. A. Button et al., "Power Failure: Why Small Sample Size Undermines Reliability of Neuroscience," *Nature Reviews Neuroscience* 14 (2013): 1–12.

10. H. F. Harlow and S. J. Suomi, "Social Recovery by Isolation Reared Monkeys," *Proceedings of the National Academy of Sciences* 68, no. 7 (1971): 1534–38.

11. G. Berkson, "Travel Vision in Infant Monkeys: Maturation Rate and Abnormal Stereotyped Behaviors," *Developmental Psychobiology* 1, no. 3 (1968): 170–74.

Chapter 3

1. H. F. Harlow, M. K. Harlow, and S. J. Suomi, "From Thought to Therapy: Lessons from a Primate Laboratory," *American Scientist* 59 (1971): 539–47.

2. See Frans DeWaal, *Peacemaking among Primates* (Cambridge, MA: Harvard University Press, 1989).

3. T. L. Beauchamp et al., *The Human Use of Animals: Case Studies in Ethical Choice* (New York: Oxford University Press, 2008).

Chapter 4

1. C. Bernard, *An Introduction to the Study of Experimental Medicine*, trans. from the French by Henry C. Green (n.p.: Henry Schuman, 1865), 100–102.

2. J. P. Gluck, M. W. Otto, and A. J. Beauchamp, "Respondent conditioning of Self-Injurious Behavior in Early Socially Deprived Rhesus Monkeys," *Journal of Abnormal Psychology* 94 (1985): 222–27.

3. Peter Singer, *Animal Liberation* (New York: Avon Books, 1975).

4. D. Blum, *The Monkey Wars* (New York: Oxford University Press, 1994), 94.

5. J. P. Gluck, H. F. Harlow, and K. A. Schiltz, "Differential Effect of Early Enrichment and Deprivation on Learning in the Rhesus Monkey (*Macaca mulatta*)," *Journal of Comparative and Physiological Psychology* 84 (1973): 598–604.

6. L. J. Kamin, *The Science and Politics of I.Q.* (New York: Lawrence Erlbaum Associates, 1974).

7. H. F. Harlow and S. J. Suomi, "Social Recovery By Isolation-Reared Monkeys," *Proceedings of the National Academy of Sciences* 68, no. 7 (1971): 1534–38.

8. T. S. Strongin, J. P. Gluck, and R. G. Frank, "Development of Social Behavior in an Adult Total Isolate Rhesus Monkey (*Macaca mulatta*)," *Journal of Autism and Childhood Schizophrenia* 7 (1978): 329–36.

Chapter 5

1. H. H. Holzgref, J. M. Everitt, and E. M. Wright, "Alpha-Chloralose as a Canine Anesthetic," *Laboratory Animal Science* 37, no. 5 (1987): 587–95.

2. S. W. Ammons, "Use of Live Animals in the Curricula of U.S. Medical Schools in 1994," *Academic Medicine* 70, no. 8 (1995): 740–43.

3. Donald Griffin, *The Question of Animal Awareness* (New York: Rockefeller University Press, 1976).

4. D. L. Cheney and R. M. Seyfarth, "Vocal Recognition in Free-Ranging Vervet Monkeys," *Animal Behavior* 28 (1980): 362–67.

5. H. K. Beecher, "Ethics and Clinical Research," *New England Journal of Medicine* 274, no. 24: 1354–60.

6. R. E. Brightman, *Grateful Prey: Rock Cree Human–Animal Relationships* (Oakland: University of California Press, 1993).

7. M. H. Lewis et al., "Early Social Isolation: Long-Term Effects of Survival and Cell Mediated Immunity," *Biological Psychiatry* 47, no. 119 (2000): 119–26; M. H. Lewis et al., "Long-Term Neurobiological Effects of Early Social Deprivation in Non-human Primates," *Biological Psychiatry* 31 (1992): 197.

8. Hans Jonas, "Philosophical Reflections on Experimenting on Human Beings," *Daedalus* 98 (1969): 219–47.

9. B. Rollin, *Animal Rights and Human Morality* (Buffalo, NY: Prometheus Books, 1981).

10. T. L. Beauchamp, "The Moral Standing of Animals," *Law, Medicine and Health Care* 1 (1992): 7–16.

11. F. B. Orlans, *In the Name of Science: Issues in Responsible Animal Experimentation* (New York: Oxford University Press, 1993).

12. D. N. Robinson, comment on animal research labs, *American Psychologist* 45, no. 11 (1990): 1269.

Chapter 6

1. Paola Calavieri and Peter Singer, eds., *The Great Ape Project: Equality beyond Humanity* (New York: St. Martin's, 1992).

2. C. Warden and H. Nissen, "An Experimental Analysis of the Obstruction Method of Measuring Animal Drives," *Journal of Comparative Psychology*, no. 8 (1928): 325–42.

3. M. A. Novak and S. J. Suomi, "Psychological Well-Being of Primates in Captivity," *American Psychologist* 43, no. 10 (1988): 765–73.

4. C. Cohen, "The Case for the Use of Animals in Biomedical Research," *New England Journal of Medicine* 315 (1986): 865–70.

5. David DeGrazia, *Taking Animals Seriously: Mental Life and Moral Status* (New York: Cambridge University Press, 1996).

6. Margaret Little, e-mail to the faculty of the Kennedy Institute, 2013.

7. Henry Beecher, "Ethics and Clinical Research," *New England Journal of Medicine* 274 (1966): 1354–60.

8. F. B. Orlans, *In the Name of Science: Issues in Responsible Animal Experimentation* (New York: Oxford University Press, 1993).

9. See J. Parascondola, "Physiology, Propaganda, and Pound Animals: Medical Research and Animal Welfare in Mid-Twentieth Century America," *Journal of the History of Medicine and Allied Sciences* 62, no. 3 (2007): 277–315.

Chapter 7

1. J. Swazey, M. Anderson, and K. Louis, "Ethical Problems in Academic Research," *American Scientist* 81, no. 6 (1993): 542–53.

2. C. Holder, "Scientist Convicted for Monkey Neglect," *Science* 214 (1981): 1218–20.

3. K. S. Guillermos, *Monkey Business: The Disturbing Case That Launched the Animal Rights Movement* (Washington, DC: National Press Books. 1993).

4. D. C. Bok, "Can Ethics Be Taught?," *Change* 8, no. 9 (1976): 26–30.

5. R. W. Samsel et al., "Cardiovascular Physiology Teaching: Computer Simulations vs. Animal Demonstrations," *Advances in Physiology Education* 11, no. 1 (1994): 36–46.

6. M. Romano, "CU Settles Suit over Dog Experiments," *Rocky Mountain News*, September 1, 1995, 4A.

7. W. M. S. Russell and R. L. Burch, *The Principles of Humane Experimental Technique* (London: Methuen, 1959).

8. Quoted in Noah Phillips, "UW Animal Research Oversight Committees Strive for Consensus," *Capital Times*, July 31, 2014.

9. T. Nagel, "What Is It Like to Be a Bat?," *Philosophical Review* 83, no. 4 (1974): 435–50.

10. Andy Clark, *Being There: Putting Brain, Body, World Together Again* (Cambridge MA: MIT Press, 1997).

11. Daniel Dennett, *Kinds of Minds: Toward an Understanding of Consciousness* (New York: Basic Books, 1991).

12. Bernard Rollin, *The Unheeded Cry: Animal Consciousness, Animal Pain, and Science* (Ames: Iowa State University Press, 1998), 144.

13. W. Russell, "On Misunderstanding Animals," *UFAW Courier* 12 (1956): 19–35.

14. W. A. Mason, "Effects of Social Interaction on Well-Being: Developmental Aspects," *Laboratory Animal Science* 41 (1991): 323–28.

15. Viktor Rheinhardt, "An Environmental Enrichment Program for Caged Rhesus Monkey at the Wisconsin Regional Primate Research Center," in *Through the Looking Glass: Issues of Psychological Well-Being in Captive Nonhuman Primates*, ed. Melinda Novak and Andrew Petto (Washington, DC: American Psychological Association, 1998), 149–59.

16. Jonathan Lear, *Radical Hope: Ethics in the Case of Cultural Devastation* (Cambridge, MA: Harvard University Press, 2006).

Chapter 8

1. J. Panskepp et al., "Declaration on Consciousness," presented at the Francis Crick Memorial Conference on Consciousness in Human and Nonhuman Animals," Churchill College, University of Cambridge, July 7, 2012.

2. J. P. Garner, "The Significance of Meaning: Why Do over 90% of Behavioral Neuroscience Results Fail to Translate to Humans, and What Can We Do to Fix It?," *ILAR Journal*, 55, no. 3 (2014): 438–56.

3. R. McManus, "Ex-Director Zerhouni Surveys Value of NIH Research," *NIH Record* 65 (2013): 1, 6, 5.

4. K. A. Phillips et al., "Why Primate Models Matter," *American Journal of Primatology* 76, no. 9 (2014): 801–27.

5. S. Wigley, "Disappearing without a Moral Trace? Rights and Compensation in Times of Emergency," *Law and Philosophy*, 28 (2009): 617–49.

6. H. K. Beecher, "Ethics and Clinical Research," *New England Journal of Medicine* 274 (1966): 1354–60.

7. National Commission for the Protection of Human Subjects in Biomedical and Behavioral Research, *The Belmont Report: Ethical Principles and Guidelines for the Protection of Human Subjects in Research*, DHEW Publication OS 78–0012 (Washington, DC: Department of Health, Education and Welfare, 1978).

8. Institute of Medicine, *Chimpanzees in Biomedical and Behavioral Research: Assessing the Necessity* (Washington, DC: National Academy Press, 2011).

Epilogue

1. John Bowlby, *Charles Darwin: A New Life* (New York: W. W. Norton, 1992), 420–21.

Index

JG = John Gluck.

Abe (family grocer), 17
aboriginal hunting practices, 208
Ahmed, Nasir, 255
air blast and shuttle box testing, 53–55
Al (lab monkey), 48–50, 51–53
Alice (infant rhesus monkey), 167
alpha-chloralose, 189–90
American Anti-Vivisection Society, 241
American Psychologist, 77
anesthesia: and cardiac physiology re-
    search on dogs, 189–90; ketamine,
    110, 113, 129, 295; NIH brain study, 206,
    207
animal cruelty: chain collars embedded in
    monkey necks, 127–29; and isolation ex-
    periments, 74; JG's childhood memories
    of witnessing, 18–20, 36–37; pain and
    distress caused by lack of surgical skill,
    150–52; "pit of despair" experiments,
    159–60
Animal Liberation (Singer), 157–59, 161,
    246
animal research: animal protocol reviews,
    260–61, 269–72; Animal Welfare Act
    of 1966, 152, 153–54; cardiac physiol-
    ogy research on dogs, 188–90; early
    childhood involvement in, 224–25; and
    ethical retraining of scientists, 13–14,

142, 143, 145, 284–86; and investigator-
    subject conflicts, 184; JG's Last Lecture
    presentation, 209–13; justifications for,
    5–9, 158–59, 195–96, 224–25, 231–34, 265;
    and moral residual, 289–91; NIH brain
    study, 202–9; nonanimal alternatives to,
    278–79. *See also* ethics
Animals and Society Institute, 242
animal welfare: *Animal Liberation* (Singer),
    157–59, 161, 246; animal protocol
    reviews, 260–61, 269–72; arguments
    against, 260–62; Cambridge Declara-
    tion, 283; concepts and costs of harming
    animals, 228–29, 259–60; discussion of
    animal ethics in bioethics presentations,
    261–65; housing requirements, 122, 230–
    31, 272–74; hunting and animal ethics,
    263–64; lack of disease as standard for,
    7, 231; and mainstream science, 283–84;
    moral status and personhood concepts,
    227–28; neurosurgery on lab rats, 166;
    on-campus debates on animal-use
    ethics, 256–60; ongoing challenges of,
    282–83; Research Ethics Service Project
    (RESP), 247, 250–56, 261, 265–67; "three
    Rs" of animal research welfare systems,
    264–65, 278–79, 289; UNM primate lab
    break-in and release of lab monkeys,

animal welfare (*cont.*)
160–62; UNM School of Medicine lab dogs, 154; and zoo administration, 165. *See also* ethics; Institutional Animal Care and Use Committees (IACUCs)

Animal Welfare Act: amendments to, 186–88; and limitations of IACUCs, 267–69; minimalist standards for, 230–31; passage of, 152, 153–54

Animal Welfare Institute (AWI), 239–41

anthropomorphism, 230–31

apology rituals, 208

Aristotle, 227, 235

associative learning and memory research, 196–97

Augustine, Saint, 4

aversive conditioning, 53–55

Bahm, Archie, 200

Balcombe, Jonathan, 241

Baltimore, David, 217–18, 251

Bass, Barbara, 32, 33

Beauchamp, Alan, 193, 203, 215

Beauchamp, Tom L., 217–18, 223, 227, 243, 247, 251, 266

Beecher, Henry, 4–5, 199, 200, 238, 289–90

behavioral and developmental psychology, isolation experiments, 73–78

behavioral science: appeal of research career in, 39–40; and cognitive ethology, 191–93; radical behaviorists, 30, 31; sanitation of research language regarding, 8, 31–32, 50, 58–59, 158

Belmont Report, 200, 201, 223, 290

Bentham, Jeremy, 227

Berkson, Gershon, 86

Bernard, Claude, 145, 148

Bernard (blacktail prairie dog), 37–39

Berridge, Craig, 268

Bessemer, Dave, 95, 112

bioethics fellowship: Animal Welfare Institute, 239–41; animal welfare organizations, 241–42; classes and seminars at, 226–27, 234, 235–36; Cohen argument on use of animals in biomedical research, 231–34; Columbia Obstruction Box, 229–30; concepts and costs of harming animals, 228–29; Exorcist Stairs cat, 248–49; Georgetown campus, 221–22, 248–49; initial group meeting, 222–23; JG's goals for, 222–23, 226, 249; Kennedy Institute of Ethics, 218–19, 234–35; mentor F. Barbara Orlans, 223–26, 238–40, 243, 247, 249, *ffig. 16*; mentor Tom L. Beauchamp, 223, 227, 243, 247; moral status and personhood, 227–28; NIH Clinical Center internship, 236–38

bird trapping experimental proposal, 210–11

Bishop, Laura, 235

"bloat syndrome," 101–2

Bok, Derek, 254

Borunda, Gilbert, x–xi, 193, 294–95

brain function and equipotentiality theory, 2

branding of cattle, 27–28

Breuning, Stephen, 251

Bruns, Steve, 69

Buddhism and external skin as "arbitrary boundary," 238

Buddy (family dog), 17

Burt, Cyril, 163–64

cages, utility-oriented, 122

Cambridge Declaration, 283

Campbell, Sam L., 34, 57–59, 64, 130

Camus, Albert, 20

cannibal sandwiches, 109

cardiac physiology research on dogs, 189–90

Carolyn (brain injury patient), 172–76

Carrol, John, 235

catheter tract infections, jugular self-infusion experiment, 150–52

cauterization, cardiac physiology research on dogs, 189–90

Center for Clinical Bioethics, 235

Charcot, Jean, 2–3

Charley (lab monkey), 194

Childress, James, 223

chimpanzees: being phased out of biomedical research, 290–91; heart transplanted into human, 3–4; at Holloman Air Force Base, 145–46; Sarah, 10; sign language acquired by Washoe, 83, 176

chronic respiratory disease (CRD) in lab rats, 165–66

Chuck (grad assistant), 127, 128

City College of New York (CCNY), 20

clinical implications of research: Carolyn (brain injury patient), 172–76; clinical fellowship sabbatical, 167–68, 169–72; as justification for primate research, 77–78

clinical psychotherapy work, 179–80, 195

cockfighting, 138

Cofer, Charles, 32

Cogan, Dennis, 34

cognition: Carolyn (brain injury patient), 172–76; cognitive development theory, 84; cognitive ecology, 191–93; embodied cognition, 270–71; food deprivation and memory testing, 146–48; JG's nature–nurture debate with Professor Wild, 162–64; and use of animals in biomedical research, 232–34

Cohen, Carl, 200, 231–34, 256–57

Colin (family dog), 131–32

college studies at City College of New York, 20–21

Collins, Francis, 291

Columbia Obstruction Box, 229–30

Comfortable Quarters for Laboratory Animals (AWI), 122, 240

compassion for animals, 13–14, 143, 145

Cork, Linda, 206, 207–8, 213–15, 226

Cross, Henry, 62

Crow Nation, 274

Dahlgren Chapel, 244–45

Darragh, Martina, 235

Darsee, John, 251

Darwin, Charles, 75, 296

data fabrication, 251–52

Daub, Guido, 117–18, 122–23

Davenport, John, 5–6, 91, 92, 94, 95, 109

Davidson, Elmer, 32–34, 40, 118, 130

Davidson, Jean, 32

Day, Phil: Friday "vet seminar" lunches, 165–66; and Holloman chimpanzees, 145, 146; NIH brain study, 206, 207; at UNM, 115–16, 119, 140, 152

deer hunting apology ritual, 208

Deets, Allyn, 86

delayed match-to-sample tasks, 125–26

Delgado, José M. R., 32

DeLuna, Susan, 110

Dennett, Daniel, 270–71

depression: Carolyn (brain injury patient), 174–75; Harry Harlow, 106–7

developmental psychology, isolation experiments, 73–78

dogs: cardiac physiology research on, 188–90, 256–60; dogfighting, 138; housing conditions at NIH, 224; Pepper and Animal Welfare Act, 153–54; at UNM School of Medicine, 154

Donna (stump-tail), ix–xii, 292

draft notice and preinduction physical examination, 87–88

Dworkin, Ronald, 227

Ebbinghaus, Herman, 84

Edwards, Jeremy (Jay), 185, 247

Edwards, Katie, 185–86, 247

eight-ball tournament, JG's UNM job interview, 99

electric shock shuttle box testing, 53

Ellis, Henry C., 170

embodied cognition, 270–71

emotions: emotional distancing, 171; and ethical retraining of scientists, 13–14, 143, 145; NIH brain study, 208–9

Enid (lab rat), 60

equipotentiality theory, 2

Estrada, Ector, x–xi, 161, 193, 207, 294

ethics: and animal consciousness, 191–93;

ethics (*cont.*)
    animals and human ethical concerns,
    1; and better scientific results, 11–12;
    discussion of in UW labs, 82–83, 104–5;
    ethical issues in death and dying, 234;
    ethically challenged clinical students,
    183–84; ethical understanding of self,
    221; and graduate research education,
    284–86; human experiments and
    research, 4–5, 162, 163, 199, 200, 201, 241,
    252, 260–61, 289–90; JG's evolution and
    transformation regarding, 9–12, 143–52,
    191, 218, 247; JG's nature–nurture debate
    with Professor Wild, 162–64; place of
    feelings in, xv–xvi; research ethics course
    at UNM, 198–202; unethical treatment
    strategies for anxiety disorders, 181–83;
    of using animals in research, 1–3, 101–2.
    See also bioethics fellowship
euthanization: B-1 rhesus male at UNM,
    113–15; Donna (stump-tail), ix–xii, 292;
    ketamine, 110, 113, 129, 295; Lala (stump-
    tail), 292–96; Millie (stump-tail),
    292–96; Pup (blind lab dog), 155–57
Exorcist Stairs cat, 248–49
experimental design: bird trapping
    experimental proposal, 210–11; Mini-
    Experiments and Micro-Methodology
    research seminar, 84–86; pain assess-
    ment and animal protocol reviews,
    269–72; "think photographically,"
    84–86, 107–8
external skin as "arbitrary boundary," 238

Faden, Ruth, 226
"false consensus effect," 214
Ferlinghetti, Lawrence, 20
Ferraro, Doug, 115, 123, 124–25, 129, 145–46,
    150, 151–52
Ferrier, Daniel, 2
food deprivation and memory testing,
    146–48
food reinforcement and isolate monkey
    training, 86

Frank, Bob, 132–36, 141–42, 167
Frankenstein (Shelley), 4
Freud, Sigmund, 20, 29, 30
Frey, Raymond, 262

Gardner, Alan, 83
Gardner, Beatrice, 83
Garner, Joseph, 283
gastric dilatation, 101–2
genetics of intelligence, 162–64
Georgetown University: campus, 221–22,
    248–49, *fig. 15*; Kennedy Institute of
    Ethics, 218–19, 234–35
Gesell, Robert, 239
Gibson, Joan, 265–66
Ginsberg, Allen, 20
Gluck, Dorothy, 15–16, 130
Gluck, John (JG): career questioning and
    self-assessment, 151–52, 164–66, 177–78,
    185, 191–96, 226; career shift to animal
    research ethics, xiii–xvii, 243, 247, 280–
    81; childhood and upbringing, 13–21;
    clinical fellowship sabbatical, 167–68,
    169–72; clinical psychotherapy work,
    179–80, 195; and Colin (family dog),
    131–32, 142, 170, 175; with F. Barbara
    Orlans, *fig. 16*; with Marigold (rescue
    horse), *fig. 18*; with Prince (family dog),
    fig. 1; and Pup (blind lab dog), 155–57;
    research career as top priority of, 130–31,
    142–43. *See also* bioethics fellowship;
    Texas Tech University; University of
    New Mexico; University of Wisconsin,
    Madison
Gluck, John (JG's father), 14–15, 21, 130,
    137–38, 163, 171
Gluck, Kim, 64–65, 94–95, 110–13, 124,
    130–32, 142
Goldberg, Ellen H., 252, 253–55
Goldstein, Doris, 234
Goltz, Frederick, 2
Gordon, William, 219–20, 242–43, 244, 247
Gorham, Dan, 144–45
Gormezano, Isidore, 196–98

graduate studies and research. *See* Texas Tech University; University of New Mexico; University of Wisconsin, Madison

Great Ape Project, The (Calvieri and Singer), 223

Great Society initiative, 163

Greenwich Village bohemian scene, 20

Greta (stump-tail), 110, 120, 140

Grice, Bob, 99, 170

Griffin, Donald, 191–93

Haquist, Bill, 149

Hardy, James, 3

Harlow, Harry F.: depression of, 106–7; ethical implications of research by, 93–94, 192, 246; history of psychology research seminar, 103–6; isolation and early learning research, 5–6, 47–48, 167; Israel surname, 139; on Kenneth Spence, 105–6; lab atmosphere and teaching style, 55–57, 92–93, 104, 204; on Lorenzian method, 104–5; as mentor to JG, 43–45, 91–92, 95, 108, 109, 121, 123–24, 133, 168–69, 185, 219, 246; Mini-Experiments and Micro-Methodology research seminar, 84–86; National Medal of Science award, 106–7; "pit of despair" experiments, 159–60; research collaboration with JG, 129, 140, 162; Singer's critique of, 157–59; on "striptease" approach to research, 107–8; support and protection of graduate students, 88–90, 92–93, 109; at UW Primate Lab (1975), *fig.* 4; on William James, 105; and Wisconsin General Test Apparatus (WGTA), 149

Harlow, Margaret K., 70–71, 90, 106

harm assessment, 229–31

Head Start Program, 163

Healy Hall, 234, 244, 245

heart transplants, chimpanzee heart transplanted into human, 3–4

Heberer, Patricia, 251

Herpes B virus, xii, 161, 293, 295

Hertz, Heinrich, 117–18

Hinton, Tom, 160, 161

history of psychology research seminar, 103–6

Holloman Air Force Base, 115, 116, 145–46

Houser, Dan, 101–2, 103

housing and animal welfare, 272–74

human–animal relations: bifurcation of human and animal worlds, 74–75; and graduate research education, 285–86; and Texas culture, 22–29

Humane Society of the United States, 241

humans: animals' role in human imagination, 1; carnal quality of curiosity and drive for knowledge, 4; experiments and research on, 4–5, 162, 163, 199, 200, 201, 241, 252, 260–61, 289–90; human twin studies, 162, 163; monkeys as models for intellectual development of, 77–78; as responsible members of the web of life, 286–89. *See also* bioethics; ethics

Human Use of Animals, The (Beauchamp, et al.), 243

hunting, 208, 263–64

impact and explosive tests at Lovelace, 126–27

In a Different Voice (Gilligan), 234

informed-consent procedures, 199–200, 236–37

Institute of Medicine report on use of chimpanzees in biomedical research, 290–91

Institutional Animal Care and Use Committees (IACUCs): conditions required for effective operation of, 279–80; effective handling of researcher misconduct, 276–78; limitations of, 267–69, 282–83; required by Animal Welfare Act, 186; research review process, xiv, 210–11, 216–18; role of veterinarians as part of, 274–76; at UNM, 187–91, 203, 255–56, 262, 267

"intellectual mind," 75–78
intelligence and nature–nurture debate, 162–64
International Science and Engineering Fair, 224
In the Name of Science (Orlans), 238–39
Irene (NIH patient's dog), 237
iron maiden apparatus, 51–52
isolation: chambers used for raising monkeys, *fig. 3*; early learning and nature–nurture question, 73–78; feeding and watering schedules for isolate monkeys, 102–3; "the isolation syndrome," 76, 85, 122; NIH brain study results, 209; operant strategies research approach, 78–79, 81–82; psychopathology of isolate monkeys, 195–96; "Social Salvation through Sibling Succorance" study, 85. *See also* rhesus monkeys
Izzy (stump-tail), 139

James, William, 104, 105
Johnson, Lyndon, 163
Johnson, Peder, 98, 99, 115, 164
Johnson, Roy, 16
Jonas, Hans, 212
Jørgenson, Henrik, 243, 247
Jørgenson, Marianne, 247
jugular self-infusion experiment, 125, 150–52
Jung, Carl, 20, 29, 30

Kaminsky, Alexis, 266
Kant, Immanuel, 227, 262
Katz, Arnie, 171, 245
Kennedy Institute of Ethics, 218–19, 234–35
ketamine, 110, 113, 129, 295
Key Bridge, 248
Kincaid, Kathleen, 31
King, Bruce, 180
Kingston, Duncan, 26–27
Kingston, Johnny, 24–25, 26
Koenig, Frances, 180

Koenig, Karl, 168, 180
Kohler, Wolfgang, 84
Kottler, Paul, 82–84, 176

lab mice: animal protocol reviews, 272; starvation of lab mice at UNM, 216–17
lab monkeys: food deprivation and memory testing, 146–48; JG's early interactions with, 6–9; "pit of despair" experiments, 159–60; primate importers and dealers, 138–39; sanitation of research language regarding, 8, 31–32, 50, 58–59, 158; "Silver Spring monkey case," 252–53, 254; understanding monkeys' point of view, 147–50. *See also* rhesus monkeys; stump-tailed macaques
lab rats: animal facility caretaker job, 59–62; biochemistry of learning and memory in, 83–84; brain surgeries on, 34–36, 166; chronic respiratory disease (CRD) in, 165–66; Columbia Obstruction Box experiments, 229–30; food deprivation prior to testing, 147; regulatory requirements for treatment of, 240–41; starvation of at UNM, 216–17; "thermal stimuli" (burn trauma) testing, 225
Lakavage, Julie, 153
Lakavage, Peter, 153
Lala (stump-tail), 292–96
Landesman, Sharon, 170
language: Harlow's descriptions of isolation experiments, 75–76; sanitation of research language, 8, 31–32, 50, 58–59; viewed as callous *versus* objective, 158
Last Lecture presentation, 209–13
Lauersdorf, Helen, 44, 69, 88
Lewis, Jon, 159–60
Lewis, Mark, 202–3, 204, 206, 214, 226
Lewis, Robert, 286–89
libido sciendi, 4
Life magazine story on lab dogs, 153
Liss, Cathy, 240

Little, Margaret, 226, 234

Logan, Frank: alcoholism of, 169–70; JG's job interview with, 96–100; JG's primate lab, 116–17, 134–35; JG's teaching career, 129–30; Logan lab, 96–97, 100, 114–15; Professor Gormezano interview and job offer, 196–98; reaction to experimental proposal requirements, 187–88

Lorenz, Konrad, 104–5

Lovelace Research Center, 118, 125–27, 128, 142

Lubbock Christian College (LCC), 21–22

Macaca arctoides. *See* stump-tailed macaques

Macaca mulatta. *See* rhesus monkeys macaques. *See* stump-tailed macaques mainstream science and animal welfare, 283–84

Manhattan Project, 4

Manny (stump-tail), 110, 120, 140

Marigold (rescue horse), *fig. 18*

Marx, Melvin, 34

Mason, William, 160

McDonald, Irene, 222

McIver, Charlene, 185–86, 191, 218, 247

McKinney, William T., 167

Mears, Clara, 168

mice, 216–17, 272

Mill, J. S., 227

Millie (stump-tail), 140, 292–96, *fig. 17*

Mini-Experiments and Micro-Methodology research seminar, 84–86

Miracle (female rhesus monkey), 103, 151

monkeys. *See* lab monkeys; rhesus monkeys; stump-tailed macaques

Monkey Wars, The (Blum), 160

Montefiore Hospital, 153–54

Moody, Kinley, 25

"moral communities," 232, 233–34

moral residual, 286–91, 296

moral status and personhood, 227–28, 257–58

Moran, Elaine, 109, 245

Morton, Herb, 249

Nagel, Thomas, 270

NASA and experimental animals, 225

Nathanson, Bernard, 222

National Cancer Society, 224–25

National Commission for the Protection of Human Subjects of Biomedical and Behavioral Research, 200, 223, 290

National Institutes of Health (NIH): brain study and closure of UNM rhesus colony, 202–9, 226; brain study and Dr. Cork's data requests, 213–15; JG's Clinical Center internship at, 236–38; Laboratory of Comparative Ethology, 204–5; lab procedures and animal housing conditions at, 223–24

Nelson, Tina, 241

New England Journal of Medicine, 4

New Mexico Board of Psychologist Examiners, 180–81, 185–86, 199

Novak, Melinda, 167, 230–31

Novum Organum (Bacon), 4

Nozick, Robert, 217–18, 227

"nuclear family" study, 70–71

Nullius in verba (on the authority of no one), 187

Oates, Charley, 23–26, 208

Oates, Halley, 24

Oates, Jay, 24

obligations *versus* rights, 232–33

O'Hair, Kevin, x, xi

Oppenheimer, J. Robert, 4

Orlans, F. Barbara, 218, 223–26, 238–40, 243, 247, 249, 266, *fig. 16*

Orth, Eleanor, 112

Oxford University Press, 239, 243

Paddy (pet bird), 17–18

Paderewski, Ignacy, 108

pain: assessment of for animal protocol reviews, 269–72; caused by lack of surgical skill, 150–52; concepts and costs of harming animals, 228–29, 259–60; female rhesus with "bloat syndrome," 101–2

"partial isolation rearing," 72

patient vulnerability and ethical clinical treatment, 181–84

Paul, Bob, 113, 114

Pavlov, Ivan, 30, 78

Pavlova, Anna, 108

Pavlovian learning and memory research, 196–97

Pavlov sling, 58

Pearce, Harold, 132

Pellegrino, Ed, 226, 235–36, 240, 245

Penny (family dog), 17, 102

Pepper (Dalmatian), 153

personhood concept, 227–28

"phantom limb" moves, 81

philos sophia and PhD degrees, 108

Piaget, Jean, 84

"pit of despair" experiments, 159–60

plagiarism, 252

Polidora, V. J. (Jim), 53, 62–63, 64, 67–68

pool tournament, author's job interview at UNM, 99

Powers, Madison, 227

Pratt, Charley, 80–81

Premack, Ann, 10

Premack, David, 10

Primate (film), 217–18

Prince (family dog), 17, fig. 1

Principles of Biomedical Ethics, The (Beauchamp and Childress), 223

Prometheus myth, 4

psychological well-being of experimental animals, 230–31

Psychologists for the Ethical Treatment of Animals (PsyEta), 241

psychology: behavioral and developmental psychology isolation experiments, 73–78; classes at Texas Tech University, 29–32; Harlow's history of psychology research seminar, 103–6; reaction of fellow students to animal research at, 36–39. See also behavioral science

psychopharmacology research, 115

Pup (blind lab dog), 155–57

Question of Animal Awareness, The (Griffin), 191

rabbit hunting, 25–26

rabbit learning and memory research, 196–98

radical behaviorists, 30, 31

Rankester, Ray, 296

rats. See lab rats

Rawls, John, 227

reciprocity and moral residual concept, 286–89

Regan, Tom, 200, 262

Research Ethics Service Project (RESP), 247, 250–56, 261, 265–67

residual debt, 286–91, 296

Rheinhardt, Viktor, 273–74

rhesus monkeys: Alice, 206; B-1 male at UNM, 97, 112–15; described, ix; E3 (male), 205–6; E7, 121; E12 (male), 81, 121; E15 (female), 81, 121, 167; E25 (female), 79–80, 207; E38 (female), 194; ethical implications of Harlow's research on, 93–94; female rhesus with "bloat syndrome," 101–2; G44 (female), 141; G49 (female), 80–81, 194, 195; at Harlow lab, 48–50, 51–53, 57, fig. 6; Harlow's isolation experiments, 75–78; J90 (male), 194–95; at Logan lab, 97; at Lovelace Research Center, 125–27; moved to NIH Laboratory of Comparative Ethology, 204–5; NIH brain study and closure of UNM rhesus colony, 202–9; "nuclear family" study, 70–71; single-cage housing of, fig. 5; at UNM primate lab, *fig. 12–14*; and Wisconsin General Test Apparatus (WGTA), 149

Richard, Mike, 165–66

Richardson, Henry, 227

"rights" and use of animals in biomedical research, 232–34

Rio Grande Zoo, 164, 193

Ripp, Chris, 69, 138

RNA and learning in rats, 83–84

Robbins, Dale, 37–39

Roberts, Laura, 266

Robinson, Daniel, 218

Robinson, Nancy M., 169

rocking behavior, 135

Rogers, Carl R., 89–90

Rollin, Bernard, 217, 271, 281

Rose (stump-tail), 139

Rosevear, Joyce, 85

Rossler, Laverne "Ross," 51, 69

Rowan, Andrew, 241

Rubaii, Safia, 259

Rush, Boyd, 3

Russell, William, 272

Rutledge, Mark, 209, 210, 211

Sackett, Gene P. (Jim): approval of JG's research plan, 79; brain study and Dr. Cork's data requests, 214–15; departure from Wisconsin, 90–91; JG's clinical fellowship, 169; as mentor to JG, 5–6, 246; and "pit of despair" experiments, 160; research collaboration with JG, 140; Sackett lab, 68–73, 82–83, 130; statistical theory and application of observational scoring systems, 134

Sandia Mountains, 96

Sarah (chimpanzee), 10

Sartre, Jean-Paul, 20

Schiltz, Ken, 44, 45, 46, 48, 69, 109

Scholes Hall, 96

Science and Politics of IQ, The (Kamin), 162–64

scopolamine research (Singh lab), 51–53

self-directed behavior, 135, 136

self-injurious behavior (SIB) experiments, 144–45

self-mutilation: B-1 rhesus male at UNM, 97, 112–13; and "the isolation syndrome," 76, 122

Senko, Monte, 67

Shapiro, Ken, 241, 242

Shelton, Steve, 206

shocking of monkeys: delayed match-to-sample tasks at Lovelace, 126; self-injurious behavior (SIB) experiments, 144–45

"shock scrambler" apparatus, 57–58

shock tubes, 127

shuttle box testing, 53–55

Silent Scream, The (film), 222

"Silver Spring monkey case," 252–53, 254

Singer, Peter, 157–59, 200, 227, 262

Singh, Sheo, 50–53

Skinner, B. F.: Harlow's history of psychology research seminar, 104; operant learning studies, 84; on predicting and controlling behavior, 30; on rejection of mentalistic explanations, 58; and Sam Campbell, 34, 57–58; *Walden Two*, 31

Skinner, Roger, 95

"small-N" mother-surrogate attachment studies, 84

Snyder, S. Bret: euthanization of Pup, 155–57; Friday "vet seminar" lunches, 164–66; JG's friendship with, 145–46, 147–48, 149–50, 166, 193; shutdown of jugular self-infusion experiment, 150–52

"social mind," 75–78

"Social Salvation through Sibling Succorance" study, 85

societal norms and "do unto others" ethical framework, 13

"son of a gun stew," 28–29

Spence, Kenneth, 95, 104, 105–6

Sproul, Jeff, 118–24

St. Mary's Hall, 221, *fig. 15*

Stevens, Christine, 239–40, 241
straight alley mazes, 97, 187
Strong, Patrick, 32
Strongin, Tim, 132–36, 167
stump-tailed macaques: described, ix;
    Donna, ix–xii, 292; at feed supply store,
    119–20, 137–38; H89 (female), 212–13;
    from Holloman Air Force Base, 127–29;
    Lala, 292–96; Millie, 292–96, fig. 17; as
    patients, 2–3; shuttle box testing, 54–55;
    at UNM primate laboratory, 136, 138–40,
    142, fig. 9–11
Sulmasy, Dan, 226–27
Summerlin, William, 251
Suomi, Stephen, 77, 85, 109, 167, 185, 204,
    205, 226, 230–31
Swaggart, Jimmy, 173
Swazey, Judith, 252, 253
symptom lists and Carolyn (brain injury
    patient), 172–76

Takemura, Kenichi, 41–42, 44
Tatum, Earl, 204–5
Taub, Edward, 253
Taylor, Carol, 235–36
Telemachus (stump-tail), x
Terman, Lewis, 139
Texas culture and human–animal relations,
    22–29
Texas Tech University: animal facility
    caretaker job, 59–62; Campbell lab, 34,
    57–62; Cogan lab, 34–36; comparative
    anatomy class at, 37–39; Davidson lab,
    32–34; decision to attend, 21–22; grad-
    uate studies at, 40–41, 57–62; harvester
    ant research, 59; psychology classes at,
    29–32
"thermal stimuli" (burn trauma) testing on
    lab rats, 225
"think photographically" experimental
    design, 84–86, 107–8
Thompson, Dennis, 254
Thompson, Wayne, 55, 67–68

"three Rs" of animal research welfare
    systems, 264–65, 278–79, 289
Titchener, Edward, 104
Tripp, Dick, 71, 72
Tuskegee syphilis studies, 200, 290

umwelt concept, 104
unconscious world, 29–30
United Kingdom animal lab practices, 224
United States Holocaust Memorial Mu-
    seum, 251
University of Nevada–Reno, 83
University of New Mexico: Animal Care
    Committee, 116, 145, 152–54, 187; B-1
    office space, 112, 115; interview at,
    94–95, 96–100; Psychology Depart-
    ment acting chair position, 213, 215,
    218–20; Psychology Department chair
    position, 219–20, 242–47; Psychology
    Department research ethics course,
    198–202; sabbatical, 218–20; School
    of Medicine lab dogs, 154; teaching
    and research career at, 129–30, 140–41;
    tenure, 166–67. See also Institutional
    Animal Care and Use Committees
    (IACUCs)
University of New Mexico primate
    laboratory: arrival of monkeys from
    Wisconsin, 120–21; break-in and release
    of monkeys, 160–62; establishment of,
    6–7, 99–100, 124, fig. 7; indoor–outdoor
    housing facility, 134–36; JG at, fig. 8; lab
    setup tasks, 100–101; rhesus monkeys at,
    123, 124–25, 132–37, 140; space in chem-
    istry department for, 117–19, 121–23;
    stump-tailed macaques at, 136, 138–40,
    142; unauthorized pain medication
    purchases, 215–16
University of Washington Regional
    Primate Research Center, 90, 91
University of Wisconsin, Madison: De-
    partment of Psychology and Harlow lab,
    89–90, 92; departure from, 109–10; JG's

graduate research at, xii–xiii, 5–6, 65–66, 108, *fig. 2*; Polidora lab, 53–55, 62–63, 64, 66–73; Primate Laboratory at, 41–43, 45–50, 55–57, 66–73; Sackett lab, 68–70, 71–73; Singh lab, 50–53

Uustall, Diann, 243, 247, 249

Veatch, Robert, 226, 234

"vertical chamber" experiments, 159–60

vervet monkey vocalizations and communication, 191–92

veterinarians: and euthanization of Pup, 155–57; Friday "vet seminar" lunches, 164–66; role on Institutional Animal Care and Use Committees (IACUCs), 274–76; and shutdown of jugular self-infusion experiment, 150–52

Vietnam War, 86–89

Vollman, Jochim, 247

Von Uexkull, Jakob, 104

Walden Two (Skinner), 31

Walters, Leroy, 226

Washoe (chimpanzee), 83, 176

Watson, John, 30, 36

Weisbard, Charley, 104

Welsome, Eileen, 253

Westinghouse Science Talent Search, 224

Wheelis, Allan, 238

Wild, Gaynor, 162–64

Wisconsin General Test Apparatus (WGTA), 46–50, 54, 149

Wisconsin Regional Primate Research Center, 5–6, 45, 82–83, 123–24

Wiseman, Frederick, 217, 218

World Medical Congress in London (1881), 2

Wright, Ted, *fig. 8*

Zerhouni, Elias, 283–84

zoo administration, 165